Power Systems Amid the 4th Industrial Revolution

RIVER PUBLISHERS SERIES IN POWER

Series Editor

MASSIMO MITOLO
Irvine Valley College,
USA

VIJAY K. SOOD
Ontario Tech University,
Canada

The "River Publishers Series in Power" is a series of comprehensive academic and professional books focussing on the theory and applications behind power generation and distribution. The series features content on energy engineering, systems and development of electrical power, looking specifically at current technology and applications.

The series serves to be a reference for academics, researchers, managers, engineers, and other professionals in related matters with power generation and distribution.

Topics included in the series are as follows:-

- Power generation;
- Energy services;
- Electrical power systems;
- Photovoltaics;
- Power distribution systems;
- Energy distribution engineering;
- Smart grid;
- Transmission line development.

For a list of other books in this series, visit www.riverpublishers.com

Power Systems Amid the 4th Industrial Revolution

Editors

Diaa-Eldin A. Mansour

Faculty of Engineering, Tanta University, Tanta, Egypt

Mohamed Gamal Hussien

Faculty of Engineering, Tanta University, Tanta, Egypt

Partha Sarathi Subudhi

Bajaj Institute of Technology, Wardha, Maharashtra, India

A. Rakesh Kumar

KPR Institute of Engineering and Technology, India

Sanjeevikumar Padmanaban

University of South-Eastern Norway, Norway

River Publishers

Routledge
Taylor & Francis Group
NEW YORK AND LONDON

Published 2024 by River Publishers
River Publishers
Alsbjergvej 10, 9260 Gistrup, Denmark
www.riverpublishers.com

Distributed exclusively by Routledge
605 Third Avenue, New York, NY 10017, USA
4 Park Square, Milton Park, Abingdon, Oxon OX14 4RN

Power Systems Amid the 4th Industrial Revolution / Diaa-Eldin A. Mansour, Mohamed Gamal Hussien, Partha Sarathi Subudhi, A. Rakesh Kumar and Sanjeevikumar Padmanaban.

Routledge is an imprint of the Taylor & Francis Group, an informa business

ISBN 978-87-7022-676-9 (hardback)
ISBN 978-87-7004-183-6 (paperback)
ISBN 978-10-0079-387-1 (online)
ISBN 978-1-003-33911-3 (ebook master)

Contents

Preface

In recent years, electrical power grids are facing several challenges. The first of these challenges is the increasing penetration of renewable energy sources to limit greenhouse gas emissions (GHE) and agree with the Paris Agreement policies. The second challenge is the continuous requirements of customers to improve the efficiency, quality, and reliability of electrical service. The third challenge is the liberalization of electricity markets from a centralized form to a decentralized one. The fourth challenge is the change of customers from pure consumers to prosumers or the emergence of new types of loading, such as electric vehicles. As a result, conventional grids cannot accommodate all these challenges. Accordingly, smart grids have emerged, especially with the great advances in information and communication technologies.

The transition to smart grids can be put into three dimensions: decentralization through integrating distributed and small power generation units, decarbonization to meet the GHG reduction target and get to the Net Zero plan, and digitalization by incorporating digital systems in electrical grids. Smart grids are not only targeting to overcome the abovementioned challenges but also targeting to add new features to the existing grids, such as advanced metering, dynamic controllability, and fault prevention. In this regard, significant efforts are exerted toward using the Industry 4.0 technologies such as the Internet of Things (IoT), artificial intelligence, digital twin, big data, machine learning, etc., to accelerate the transition to smart grids. This book discusses the operation of power systems amid the 4th Industrial Revolution and the value of Industry 4.0 technologies to grid operators and end customers.

Chapter 1 introduces the climate change and energy sector, highlighting the transition of electrical power grids to smart grids.

Chapter 2 provides an overview of the IoT concept and its various forms in different parts of the electrical power system within the smart grid context.

Chapter 3 describes cyber–physical power systems and discusses a variety of modeling, simulation, and analytic tools together with various forms of cyber-attacks and cybersecurity actions for current cyber–physical power systems.

Chapter 4 presents recent advances in IoT and machine learning to troubleshoot smart energy systems, including fault diagnosis and cybersecurity for several power systems components such as gas-insulated switchgears, power transformers, and electrical machines.

Chapter 5 presents a comprehensive review of the applications of the IoT and digital twin technologies in energy management in electrical power systems by optimizing, controlling, and monitoring energy consumption.

Chapter 6 outlines a comprehensive analysis and information about the applications of the 4th Industrial Revolution technologies for renewable energy operation and control.

List of Contributors

Amer, Mohammed, *Department of Mechanical Engineering, Palestine Technical University – Kadoorie, Palestine*

Bamisile, Olusola, *School of Mechanical and Electrical Engineering, University of Electronic Science and Technology of China, China; Sichuan Industrial Intelligent and Applications Lab, Chengdu University of Technology, China*

Darwish, Mohamed M. F., *Department of Electrical Engineering, Faculty of Engineering at Shoubra, Benha University, Egypt; Department of Electrical Engineering and Automation, School of Electrical Engineering, Aalto University, Finland*

Deepa, T., *School of Electrical Engineering, Vellore Institute of Technology, India*

Eid, Ahmad, *Department of Electrical Engineering, Faculty of Engineering, Aswan University, Egypt; Department of Electrical Engineering, College of Engineering, Qassim University, Saudi Arabia*

El-Kishky, H., *Electrical Engineering Department, The University of Texas at Tyler, USA*

Elsisi, Mahmoud, *Department of Electrical Engineering, National Kaohsiung University of Science and Technology, Taiwan; Department of Electrical Engineering, Faculty of Engineering at Shoubra, Benha University, Egypt*

Han, Yang, *School of Mechanical and Electrical Engineering, University of Electronic Science and Technology of China, China;*

Huang, Qi, *School of Mechanical and Electrical Engineering, University of Electronic Science and Technology of China, China; Sichuan Industrial Intelligent and Applications Lab, Chengdu University of Technology, China*

Hussien, Mohamed Gamal, *Department of Electrical Power and Machines Engineering, Faculty of Engineering, Tanta University, Egypt*

Lehtonen, Matti, *Department of Electrical Engineering and Automation, School of Electrical Engineering, Aalto University, Finland*

Mahmoud, Karar, *Department of Electrical Engineering and Automation, School of Electrical Engineering, Aalto University, Finland; Department of Electrical Engineering, Faculty of Engineering, Aswan University, Egypt*

Mansour, Diaa-Eldin A., *Department of Electrical Power Engineering, Faculty of Engineering, Egypt-Japan University of Science and Technology (E-JUST), Egypt; Department of Electrical Power and Machines Engineering, Faculty of Engineering, Tanta University, Egypt*

Numair, Mohamed, *Department of Electrical Engineering, KU Leuven, Belgium*

Ramadan, R., *School of Mechanical and Electrical Engineering, University of Electronic Science and Technology of China, China; Department of Electrical Power and Machines Engineering, Faculty of Engineering, Tanta University, Egypt*

Sanjeevikumar, P., *Department of Electrical Engineering, IT and Cybernetic, University of South-Eastern Norway, Norway*

Selvasundar, K., *School of Electrical Engineering, Vellore Institute of Technology, India*

Tran, Minh-Quang, *Department of Mechanical Engineering, TUETECH University, Vietnam; Department of Mechanical Engineering, Thai Nguyen University of Technology, Vietnam*

Vu, Viet Q., *Department of Mechanical Engineering, Thai Nguyen University of Technology, Vietnam*

Zalhaf, Amr S., *School of Mechanical and Electrical Engineering, University of Electronic Science and Technology of China, China; Department of Electrical Power and Machines Engineering, Faculty of Engineering, Tanta University, Egypt*

List of Figures

List of Tables

List of Abbreviations

4G	Fourth-generation
5G	Fifth-generation
ADC	Analog-to-digital converter
AEP	Application enablement platform
AI	Artificial intelligence
ALM	Appliance load monitoring
AMI	Advanced metering infrastructure
BCS	Best compromise solution
BD	Big data
CNN	Convolutional neural network
CPPS	Cyber–physical power system
CPS	Cyber–physical system
DER	Distributed energy resource
DG	Distributed generation
DM	Decision-maker
DSO	Distribution system operator
DT	Decision tree
EID	Electrode/Insulator delamination
EMS	Energy management system
ESS	Energy storage system
EV	Electric vehicle
FACTS	Flexible alternating current transmission system
FDD	Fault detection and diagnosis
FDM	Fuzzy-based decision-making
FIT	Feed-in tariff
FSM	Finite state machine
GHE	Greenhouse gas emission
GHG	Greenhouse gas
GIS	Gas-insulated switchgear
HC	Hosting capacity
HED	High-energy discharge
HT	High thermal

HTTP	Hypertext Transfer Protocol
HVAC	Heating, lighting, ventilation, and air conditioning
HVDC	High-voltage direct current
ICT	Information communication technology
IEA	International Energy Agency
IETF	Internet Engineering Task Force
ILM	Intrusive load monitoring
IoS	Internet of Systems
IoT	Internet of Things
IoTEP	IoT energy platform
IPSec	Internet Protocol Security
IR	Industrial Revolution
LED	Low-energy discharge
LoRa	Long range
LSTM	Long short-term memory
LT	Low thermal
MAC	Medium access control
ML	Machine learning
MOBO	Multi-objective bonobo optimizer
MOO	Multi-objective optimization
MPG	Metal particles within the gas gap
MPS	Metal particles adhering to the spacer surface
MQTT	Message Queuing Telemetry Transport
MT	Medium thermal
NDA	Non-disclosure agreement
NI	National Instruments
NILM	Non-intrusive load monitoring
NOTTLS	NO tunneled transport layer security
OLTC	On-load tap changer
OPF	Optimal power factor
OSPF	Open Shortest Path First
PD	Partial discharge
PHEV	Plug-in hybrid electric vehicle
POF	Pareto optimal front
PV	Photovoltaic
RES	Renewable energy source
RF	Random forest
SCADA	Supervisory control and data acquisition
SNMP	Standard Network Management Protocol
SOFC	Solid oxide fuel cell

SSL	Secure Socket Layer
TLS	Transport Layer Security
TOPSIS	Technique for order of preference by similarity to ideal solution
TTLS	Tunneled transport layer security
UPF	Unity power factor
USM	User-based Protection Model
VPN	Virtual private network
WSAN	Wireless sensors and actuators network
XGBoost	eXtreme gradient boosting

1

Introduction

Diaa-Eldin A. Mansour[1,2], Mohamed Numair[3], Mohamed Gamal Hussien[2], and P. Sanjeevikumar[4]

[1]Department of Electrical Power Engineering, Faculty of Engineering, Egypt-Japan University of Science and Technology (E-JUST), Egypt
[2]Department of Electrical Power and Machines Engineering, Faculty of Engineering, Tanta University, Egypt
[3]Department of Electrical Engineering, KU Leuven, Belgium
[4]Department of Electrical Engineering, IT and Cybernetic, University of South-Eastern Norway, Norway

1.1 Climate Change and Energy Sector

A rapid growth in population and industry has been noticed in the last decades. This growth has increased global electricity consumption to meet several residential, commercial, and industrial requirements. This increase results in gas emissions with a negative impact on global warming. Global warming has many consequences, including climatic changes, floods, and severe storms. The power energy production sector produces about 75% of the total CO_2 emissions worldwide [1] due to the major dependence of energy production on conventional sources that use fossil fuels. So, many efforts are exerted toward reducing these emissions by enhancing the infrastructure of the energy production sector or by saving energy across various consumption sectors.

The Paris Agreement set a legally binding international treaty on climate change between 196 parties with a common goal of keeping the global warming levels at 1.5°C compared to pre-industrial levels [2]. This agreement pushed worldwide policies to set governing rules on fossil fuels and limit greenhouse gas (GHG) emissions. In addition, significant efforts are exerted toward promoting renewable energy generation with incentivized feed-in tariff (FIT) programs. These incentives allow individuals, businesses, and investors to participate in the electricity generation pool through renewable energy sources (RESs). These incentives are risk-free investment, easier loans, and tax exemptions. These programs effectively achieve the promised RES percentage of the total generation. It is illustrated in [3] that RESs will be able to provide over 3000 times the current global energy demand. The share of renewable energy sources in the total electrical generation is expected to rise from 20% in 2010 to 31% in 2035.

According to the 2021 International Energy Agency (IEA) renewables report [4], the average annual addition of renewable electricity capacity globally would be 305 GW per year in the period between 2021 and 2026. This prediction assumes an increase of 60% compared to the historical rate. To achieve that rate and keep up to the Net Zero by 2050 promise, governments worldwide need to address some technical, social, and regulatory challenges surrounding the addition of renewables.

1.2 Factors Limiting Clean Energy Integration

Most of the proposed solutions for reducing GHG emissions depend on altering the generation structure from traditional power plants to cleaner RESs such as photovoltaics (PVs) and wind power. Such RESs convert natural resources into electrical power, but generated power inherits the stochastic nature of its resource. Another issue with RESs is their generation profiles with peaks that do not match the demand peaks, leading to the curtailment of generation or what is commonly known as the duck curve. These are not the only issues with the mass integration of RESs. One concern is the infrastructure of current distribution networks, which are built to carry energy in one direction, from generation to consumption. Since many consumers are becoming prosumers, i.e., having rooftop PV and producing electricity to the distribution network, numerous instability issues have risen on the low voltage side of the distribution networks [5]. Another concern includes the lack of investor confidence due to frequent policy changes. Moreover, the social acceptance of significant renewable energy projects caused delays for these projects.

It can be concluded that the current electrical power grids have some limitations to the maximum capacity of RESs to be added to the total capacity. These limitations and their details are coined in what is known as the hosting capacity (HC) of RESs [6]. Factors governing HC include power quality issues, stability constraints, overvoltage issues, and malfunction of protection equipment. These factors hold the addition of RESs at lateral points of the electrical grid. These limitations resulted in the suspension of some government programs. In addition to the HC of renewables, the distribution networks are facing further challenges when connecting electric vehicles (EVs) to the grid [7], a challenge expected to get more complicated.

1.3 Electrical Power Grids Transition and Smart Grids

The previous discussion clarifies that the current electrical power grids are reaching their end of life, and a paradigm shift will be witnessed in the upcoming decade. Features of this new power grid are bidirectional power flow, introducing information communication technology (ICT), and using Industry 4.0 technologies in the future smart grids.

The smart grid concept was introduced years ago [8], but the practical implementation into power grids was delayed as the ICT infrastructure was not ready [9]. Great efforts have been taking place in many countries toward the practical implementation of smart grids with the release of millions of smart meters. The smart grid features are summarized in Figure 1.1. Smart grids will benefit from Industry 4 technologies to achieve multi-layer monitoring, controllability through an advanced metering infrastructure (AMI), and higher integration of RESs and EVs to the electrical grid. Moreover, the advanced ICT infrastructure and machine learning will enable smart grids to implement new electricity market schemes and active consumer participation.

From the previous discussion, the transition to the smart grid can be put into three dimensions of future smart grids, as follows:

1. *Decentralization* means integrating distributed and small power generation units, mostly rooftop PV plants close to the consumer side, instead of the traditional centralized thermal power plants that power the whole network.

2. *Decarbonization* describes the strategic efforts in meeting the GHG reduction target to abide by the Paris Agreement agreed on levels and get to the Net Zero target.

3. *Digitalization* incorporates digital systems in electrical systems, starting with digital sensors, meters, and protection systems. It also refers to the full utilization of the ICT network and AMI to achieve technical value, accelerate decarbonization, and overcome current grid challenges. Smart Grid 2.0 performs the previous multi-layered operation through Industry 4.0 technologies such as Internet of Things (IoT), artificial intelligence (AI), and digital twin models.

It can be said that smart grids use advanced technology to overcome existing challenges and utilize new features to extract additional values. In addition,

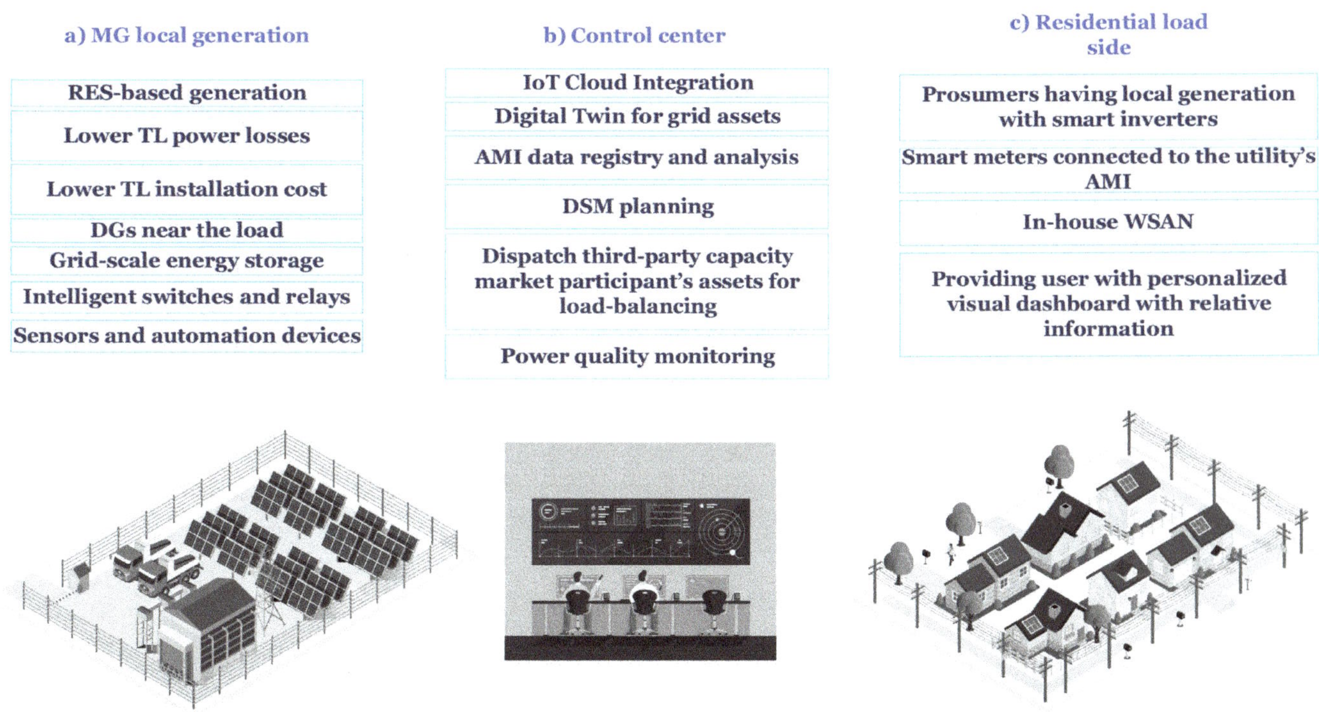

Figure 1.1 Smart grid features at (a) generation side, (b) control center, and (c) residential load side.

smart grids have other objectives, such as advanced metering, dynamic controllability, and fault prevention. Industry 4.0 technologies can enable more advanced features that can add value to all parties in the grid and achieve a more holistic efficiency increase for the entire system via accomplishing business goals and realizing technical requirements. This book discusses the operation of power systems amid the 4th industrial revolution and the value of industry 4.0 technologies to grid operators and end customers.

References

[1] R. M. Elavarasan, G. Shafiullah, S. Padmanaban, N. M. Kumar, A. Annam, A. M. Vetrichelvan, *et al.*, "A comprehensive review on renewable energy development, challenges, and policies of leading Indian states with an international perspective," *IEEE Access,* vol. 8, pp. 74432–74457, 2020.

[2] R. S. Dimitrov, "The Paris Agreement on Climate Change: Behind Closed Doors," *Global Environmental Politics,* vol. 16, no. 3, pp. 1–11, 2016.

[3] O. Ellabban, H. Abu-Rub, and F. Blaabjerg, "Renewable energy resources: Current status, future prospects and their enabling technology," *Renewable and sustainable energy reviews,* vol. 39, pp. 748–764, 2014.

[4] IEA, Renewables 2021: Analysis and forecasts to 2026, 2021. (accessed: 25/12/2023). [Online]. Available: https://www.iea.org/reports/renewables-2021

[5] G.-H. Gwon, C.-H. Kim, Y.-S. Oh, C.-H. Noh, T.-H. Jung, and J. Han, "Mitigation of voltage unbalance by using static load transfer switch in bipolar low voltage DC distribution system," *International Journal of Electrical Power & Energy Systems,* vol. 90, pp. 158–167, 2017.

[6] S. M. Ismael, S. H. E. Abdel Aleem, A. Y. Abdelaziz, and A. F. Zobaa, "State-of-the-art of hosting capacity in modern power systems with distributed generation," *Renewable Energy,* vol. 130, pp. 1002–1020, 2019.

[7] M. Nour, J. P. Chaves-Ávila, G. Magdy, and Á. Sánchez-Miralles, "Review of Positive and Negative Impacts of Electric Vehicles Charging on Electric Power Systems," *Energies,* vol. 13, no. 18, p. 4675, 2020.

[8] X. Yu, C. Cecati, T. Dillon, and M. Simões, "The New Frontier of Smart Grids," *IEEE Industrial Electronics Magazine,* vol. 5, no. 3, pp. 49–63, 2011.

[9] D. B. Avancini, J. J. P. C. Rodrigues, S. G. B. Martins, R. A. L. Rabêlo, J. Al-Muhtadi, and P. Solic, "Energy meters evolution in smart grids: A review," *Journal of Cleaner Production,* vol. 217, pp. 702–715, 2019.

2

Infrastructure for the 4th Industrial Revolution Technologies

Mohamed Numair[1], Diaa-Eldin A. Mansour[2,3], and Mohamed Gamal Hussien[3]

[1]Department of Electrical Engineering, KU Leuven, Belgium
[2]Department of Electrical Power Engineering, Faculty of Engineering, Egypt-Japan University of Science and Technology (E-JUST), Egypt
[3]Department of Electrical Power and Machines Engineering, Faculty of Engineering, Tanta University, Egypt

Abstract

The increasing demand for electrical power has led to the development of several solutions in recent years. Among them, the use of renewable energy sources has shown great promise for meeting the growing demand while reducing carbon emissions. However, electrical power systems are subject to unpredictable generation and load variations, which require dedicated control strategies. Traditional control and operation centers have been introduced in the literature, but they are heavily dependent on system parameters or rely on legacy controllers that are inefficient due to the lack of data required for optimal operation.

The advent of the smart grid concept and the Internet of Things (IoT) technology offers a potential solution for these challenges by enabling efficient data collection, storage, and analysis to accurately predict generation and control load demand. The implementation of an IoT cloud-based control strategy offers the possibility of better energy management across the electrical power system.

This chapter provides an overview of the IoT concept and its various forms in different parts of the electrical power system. It then details the advantages of using an IoT cloud-based control strategy over traditional control and operation centers. This chapter also highlights the potential of IoT

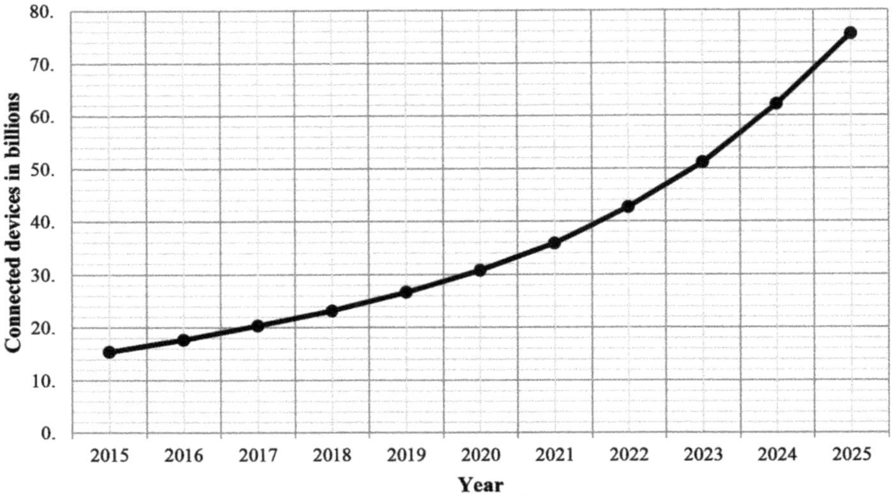

Figure 2.1 The predicted growth of IoT-connected devices over time [3].

applications for modern power system technologies to provide efficient solutions for the management and control of electrical power systems in the smart grid context.

2.1 Introduction

Machine-to-machine communication dates back to the 19th century, as documented in the literature [1]. However, it was not until the development of the internet by the Defence Advanced Research Projects Agency in 1962 that this vision became a reality [2]. In the subsequent decades, the internet has undergone significant commercialization and expansion, connecting millions of computers and, more recently, an even greater number of mobile devices. The exponential growth of connected devices to the internet is illustrated in Figure 2.1.

The Internet of Things (IoT) refers to the interconnection of physical devices for the purpose of exchanging data and information over the internet. The IoT system is composed of various technologies, including wireless communication technologies, embedded systems, wireless sensor networks, GPS, control systems, and others, which collectively enable its operation.

IoT is currently revolutionizing several industries, including the electrical power system. The integration of IoT devices into power system applications can bring numerous benefits, such as improving energy efficiency, enhancing system reliability, and enabling demand response.

One potential application of IoT in the power system is the implementation of smart grids. Smart grids are advanced electrical grids that incorporate IoT devices and technologies to improve the efficiency and reliability of power generation, transmission, and distribution [4]–[7]. IoT sensors and devices can be used to collect data on power system performance and optimize its operation in real time. For example, IoT-enabled sensors can monitor the status of power transformers and detect potential failures, enabling proactive maintenance and reducing downtime [8].

Another potential application of IoT in the power system is demand response. Demand response refers to the ability of consumers to adjust their energy consumption based on real-time information on electricity prices and system conditions. IoT devices can enable automated demand response by providing consumers with real-time information on energy prices and usage, allowing them to adjust their consumption patterns accordingly. This can lead to significant energy savings and reduce stress on the power system during peak demand periods [9]–[11].

Furthermore, the integration of IoT devices into power system applications can improve the monitoring and management of distributed energy resources (DERs). DERs, such as solar panels and wind turbines, generate power locally and feed it into the grid. IoT-enabled sensors can monitor the performance of DERs and adjust their output based on system needs, improving the stability and reliability of the power system.

The power sector's growing interest in the Internet of Things (IoT) stems from its potential to bring about significant benefits. Governments and utilities worldwide have invested in IoT-enabled smart grid technologies to enhance the efficiency and reliability of their power systems. As the technology advances and becomes more accessible, the potential of IoT to revolutionize the power sector is poised to increase. The research community has also shown a keen interest in this area, exploring the communication capabilities of an IoT infrastructure, leveraging cloud computation to optimize energy management algorithms, and examining the efficiency gains from network asset and load control [12].

As the number of IoT applications in power systems continues to grow, there is an increasing need for a common infrastructure design or blueprint that can enable modularity and interoperability between different smart grid solutions. This chapter attempts to provide that blueprint, which would help ensure that various IoT solutions can work seamlessly together, regardless of their origin or function. To achieve this, it is essential to understand the main components of an IoT system and how they interact with each other. This includes sensors, communication networks, data processing and storage, and application interfaces. A clear understanding of the value chain is also important, as it helps to identify key stakeholders and their roles in the development and deployment of IoT solutions. Ultimately, a common IoT infrastructure blueprint can help streamline the development process and promote collaboration across industry and academia, leading to faster innovation and greater adoption of IoT technologies.

The remainder of this chapter is structured as follows. Section 2.2 presents the generalized IoT value chain structure with a detailed description of each item in the IoT value chain, including potential variations and alternatives. Section 2.3 establishes the relationship between the IoT value chain elements and each of the power system components. Finally, Section 2.4 discusses some of the recent IoT applications for low-carbon technologies and advanced operation of the modern power system.

2.2 The Generalized IoT Value Chain

The IoT value chain is consistent across industries and applications, following a general structure referred to as the IoT value chain. This chain involves communication between an endpoint and the cloud, where most computations occur. The interaction between these entities provides value within each application, hence the name.

The IoT value chain can be broadly categorized into three main components, each of which consists of numerous combinations and variations. A high-level overview of the IoT value chain begins with the device or thing being monitored or controlled. This could be a power system equipment monitored or controlled through a sensor or an actuator, respectively.

The second component of the IoT value chain is the communication layer, which facilitates the transmission of data from the device to the cloud-based application enablement platform (AEP), where data analysis, computation, storage, and visualization occur. The communication layer can take various forms, such as cellular networks, Wi-Fi, Bluetooth, or even satellite communication, depending on the specific application and use case.

The third component, the cloud-based AEP, provides the necessary computational power and storage capabilities for IoT systems. It hosts the necessary algorithms for data analysis and processing and provides a visualization interface for end-users to interpret and utilize the information generated by the system.

Each component of the IoT value chain can be further subdivided and tailored to specific use cases. For instance, the device layer can consist of sensors, actuators, or both, depending on the application requirements. The communication layer can use different protocols and network types depending on the geographic location and connectivity requirements. The AEP can utilize various cloud-based services and deployment models, such as public or private cloud, to meet specific processing and storage needs.

Understanding the components of the IoT value chain is crucial in designing and implementing an IoT application for any system. Careful consideration of the specific requirements of each component can lead to a system that meets the needs of end-users while providing the necessary insights and analytics for informed decision-making.

2.2.1 The device layer

The first layer of the IoT value chain is the thing, known as the device or physical layer, which is responsible for determining the specific application of the IoT system. For instance, an IoT system designed for the healthcare industry could include a heart rate sensor, whereas an IoT system for the electrical power system might include a smart meter, transformer, cable, or energy storage device. The precise nature of IoT technology's application in the power system can vary depending on factors such as the ultimate beneficiary of the system's insights or control actions and the system's overall purpose.

It is important to distinguish between the asset being monitored or controlled and the sensors or actuators connected to it. For instance, an on-load tap changer (OLTC) transformer might have both current and voltage sensors to monitor its operation as well as a motor actuator to change the transformer taps in response to external signals. Hardware requirements for an IoT-connected asset typically include a sensor, actuator, micro-controller unit, memory, and gateway. Proper functioning of the hardware requires appropriate software, such as firmware or a real-time operating system, that can control the embedded hardware.

In addition to the basic software required to control the hardware, an application program may be necessary to provide users with an interface for monitoring and controlling local actions. Depending on the system's complexity, this software layer can range from a simple set of command-line tools to a more advanced graphical user interface. Furthermore, an IoT-connected asset may also require additional software components, such as

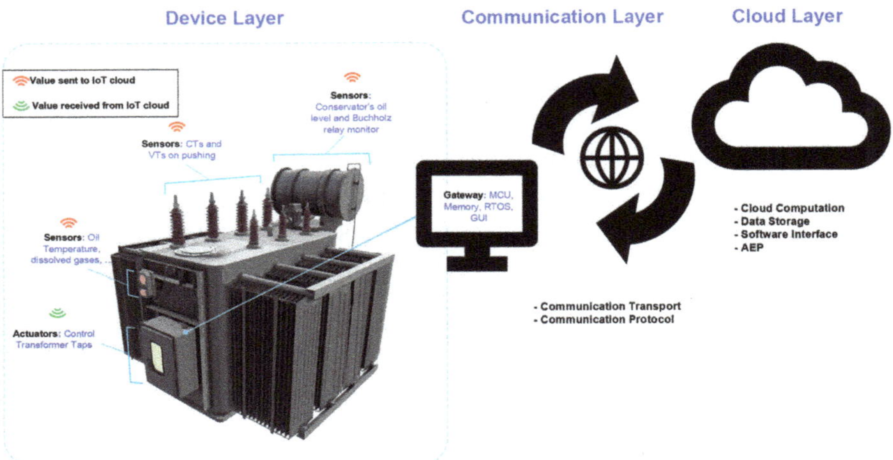

Figure 2.2 The connectivity layout for an IoT-connected OLTC transformer.

an edge application, for aggregating measured data or performing some data pre-processing locally before transmitting it to the cloud. These considerations are essential to successfully designing and implementing IoT systems across a wide range of applications.

Figure 2.2 provides a general outline of IoT connectivity for an OLTC transformer. It shows the different layers of the IoT value chain, with the device being the transformer itself along with its embedded wireless sensors and actuators network (WSAN). The device is then connected to the communication layer through a gateway. This allows for real-time monitoring of the operations of the transformer, which can capture additional insights that can be utilized to optimize its efficiency, reduce its energy consumption, and improve its performance.

To make a device IoT-ready, it needs to meet certain hardware and software requirements. The hardware components required include sensors, actuators, CPUs, gateways, and memory. These components work together to allow the device to collect and process data, creating a reliable and efficient network that supports IoT capabilities. Additionally, the device needs software components such as firmware and application programs to process and communicate data to other devices and cloud platforms. These components should be designed to work together seamlessly to ensure that the device operates effectively within the IoT ecosystem. Table 2.1 summarizes these hardware and software requirements, providing a useful guide for developers looking to create IoT-ready devices.

Table 2.1 Hardware and software requirements for developing an IoT ready device.

Hardware requirements	***Sensors***	For the purpose of our discussion, sensors can be defined as embedded system modules that convert environmental actions, events, or status into a standardized electronic/digital signal to be processed inside a CPU.
	Actuators	These devices convert electronic/digital signals into a certain action; this can be as simple as sending a signal to a motor to open or close a valve.
	CPU	Short for central processing unit, which is typically a microchip inside which arithmetic and logic operations are made. It receives all signals from different sensors and sends signals for the actuators to perform the required actions.
	Gateway	A gateway creates a local network, usually a wireless network, to connect different devices, sensors, and actuators with the CPU. It also provides the means for the local network with the CPU to connect to the internet.
	Memory	It is essential for any computing system to have some sort of memory. The size and how long the data are kept in the memory are fully dependent on the application.
Software requirements	***A firmware***	Some kind of firmware, namely a real-time operating system, is required to be installed on the embedded hardware in order to provide the development environment for application programs.
	Application programs	This is the layer at which an application program is written to run on top of the operating system, to perform certain algorithms or execute certain operations that serve the goal of the local embedded system and the objective and value extraction of the whole IoT system.

2.2.2 The communication layer

In order to facilitate signal transport between IoT devices and the cloud, a communication medium is essential. This medium can be divided into two parts. The first is the WSAN, which is a short-range local network linking IoT devices and the local gateway unit within a range of 10–100 meters. The WSAN is responsible for establishing a dependable connection between the central processing unit and various sensors and actuators. Because many sensing devices rely on batteries, communication chips with low power

Table 2.2 Comparison between the common WSAN communication protocols [13], [14].

	Bluetooth	**ZigBee**	**Wi-Fi**
IEEE specification	IEEE 802.15.1	IEEE 802.15.4	IEEE 802.11b
Frequency band	2.4 GHz	868/915 MHz; 2.4 GHz	2.4 GHz; 5GHz
Maximum signal rate	1 Mbps	250 Kbps	54 Mbps
Type of module	HC-05	XBee Series 1	Espressif modules
Sleeping mode	9 μA	12 μA	30 μA
Awake mode	35 mA	50 mA	245 mA
Transmitting mode	39 mA	52 mA	251 mA
Receiving mode	37 mA	54 mA	248 mA
Power supply	3.3 V	3.3 V	5 V
Maximum range	~9 m	~60 m	~70 m

requirements are chosen for the WSAN. Wi-Fi, ZigBee, Bluetooth, and Bluetooth low energy are commonly used as transport protocols for WSAN communication.

Table 2.2 presents a thorough comparison of Bluetooth, ZigBee, and Wi-Fi, which are all popular short-range communication protocols used in WSANs. Table 2.2 compares important aspects that would concern IoT developers when designing such networks, such as maximum range, data rate, and current consumption for different operational modes. Given that many WSAN devices are battery-powered, an efficient communication protocol is crucial, even if it means sacrificing higher data rates.

The second part of the communication medium is used for long-distance communication between devices, systems, and the cloud platform of the IoT AEP. Protocols used in long-distance communication must balance high data transfer speed with low energy consumption due to the large physical distance between systems. Low-power communication protocols such as LoRa, LoRaWAN, and SigFox have been introduced to support low-data-rate IoT communications. While higher data rates are achievable with cellular networks, they also result in higher energy usage, making them less than ideal for IoT applications.

Recent improvements in cellular communication technology have resulted in faster data transfer speeds and improved energy efficiency. The 5G technology has enabled IoT devices to communicate more quickly and reliably, supporting a larger number of devices and transmission needs. 6G, the next generation of cellular technology, is being developed to bring even faster data rates, lower latency, and wider network coverage, further enhancing the capabilities of IoT devices and networks.

Table 2.3 provides a comparison of four long-range IoT transport options, including LoRa, LoRaWAN, SigFox, and Cellular transport. The

Table 2.3 Comparison between common long-range transports [15], [16].

Transport	Long range (LoRa)	LoRaWAN	SigFox	Cellular
Range	15–30 km	5–15 km	10–15 km	1–50 km
Frequency	433–928 MHz	868–915 MHz	868 MHz	800–1900 MHz
Bandwidth	0.3–50 Kbps	0.3–50 Kbps	100–600 bps	4G: up to 100 Mbps 5G: 1–20 Gbps
Power usage	Low	Low	Low	High

key factors considered when comparing these transport modes are the range, frequency, bandwidth, and power usage. The range is vital in selecting the appropriate transport mode, as it impacts the distance over which data can be transmitted. The frequency and bandwidth are also crucial factors as they determine the data rate and capacity of the transport mode. Generally, a higher bandwidth would have a lower range and higher power usage. Conversely, a lower bandwidth will have a longer range and lower power usage.

The data of Table 2.3 provide valuable insights into the advantages and disadvantages of each transport mode and helps designers make informed decisions about which transport mode to choose for their IoT applications. A more detailed discussion of each of these technologies is provided below:

2.2.2.1 LoRa

LoRa is a proprietary technology, and its chips are only manufactured by SemTech. However, it is free to use, and there is no quota limit on messages, unlike other commercial low-power wide-area network solutions. It can be used at any frequency, but it is commonly used at 315, 433, 868, and 900 MHz of the industrial, scientific, and medical bands. This frequency choice feature, along with its relatively low power to its range and the free quota at a fair rate of 50 Kbps, makes LoRa stand out.

On the other hand, it has a simple structure as it is only meant for sending and receiving packets of data. The additional networking features must be set up by the developers themselves. A more convenient option is using LoRaWAN on top of LoRa modules, as it provides more networking options. For relatively large areas, such as in our case, microgrids, LoRa is a strong candidate for networking.

2.2.2.2 SigFox

SigFox is a pre-existing network managed by the company of the same name. The SigFox chips can only be used in cities or locations where the network is already established. This solution provides ease of mind, as developers do not have to worry about networking, managing gateways, and range. It also

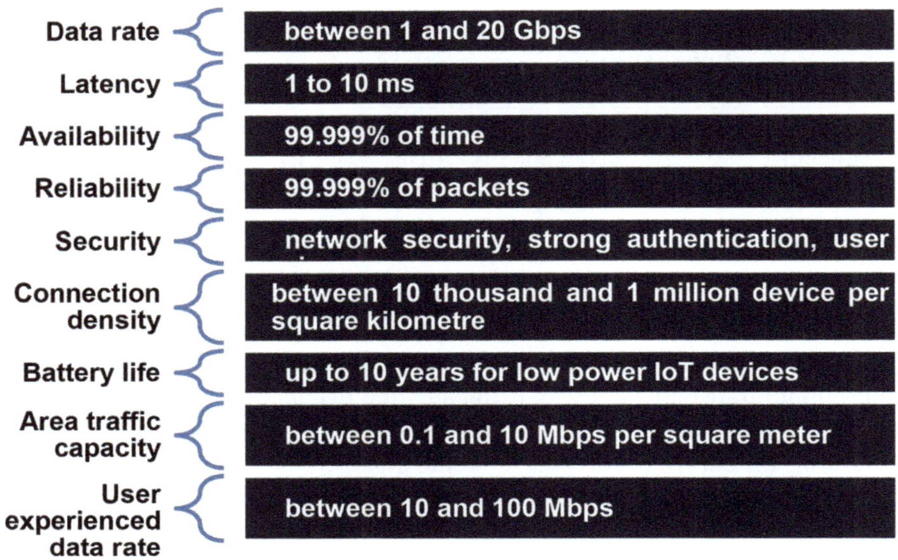

Figure 2.3 The 5G-IoT enabling features for power system applications.

excels in requiring very little power. However, there are some drawbacks to consider. First, it is a paid subscription service. Second, its speed is limited to 100 bps, making it unsuitable for high-velocity data. The most significant constraint is its limited upload bandwidth of only 140 12-byte messages per day and download bandwidth of only four 8-byte messages per day. Another consideration when attempting to use SigFox is its availability in the location where the IoT network is being deployed.

2.2.2.3 Cellular

Nowadays, almost everyone has a cell phone with a SIM card that allows them to call, send messages, and use the internet from anywhere. For inter-country communications, cellular technology is the obvious choice. Although not new, cellular technology has undergone numerous improvements from the legacy 2G/GSM to the current 5G.

The existing 4G technology has been used in many IoT applications, resulting in the term narrowband IoT. However, more is expected from the fifth generation of cellular networks, which was developed with IoT in mind and is known as 5G-IoT architecture. For power systems, 5G has a lot of potential due to its high speeds, superior reliability, and negligible latency. Figure 2.3 showcases the 5G-IoT features; these features make it highly suitable for smart grid applications, as electrical systems require sub-second controls, and unreliable communication or delays cannot be tolerated [17]. This is especially important as it enables these high speeds for long ranges, enabling applications such as wide area control or wide area protection. Nevertheless, cellular technology requires more power than other means of transport mentioned previously, and subscription costs to use different cellular networks are much higher, especially for international use.

2.2.3 The cloud layer

The cloud layer of the IoT value chain serves as a foundation for the AEP, providing the essential infrastructure for device connectivity and data management. Data collected from IoT devices is stored, processed, and analyzed in the cloud using various technologies such as big data platforms and machine learning algorithms. AEPs leverage this cloud infrastructure to provide businesses with device management, data storage and analysis, and visual dashboards for monitoring and controlling IoT systems. The cloud layer also facilitates secure communication between devices and applications, enabling real-time data transmission and remote device management. Together, the cloud layer and AEP play a crucial role in enabling businesses to derive valuable insights and make data-driven decisions from their IoT systems.

Various industries have implemented IoT systems with distinct business, technical, or economic objectives, leading to differing data storage, analysis, and computation methods among applications. To achieve these business goals, a variety of AEPs have been developed for each application. An AEP is a platform that facilitates the development and deployment of IoT applications by providing device management, data storage and analysis, and visual dashboards for monitoring and controlling the IoT system. A significant benefit of AEPs is that domain experts do not need to handle IoT and cloud infrastructure development, instead focusing on using the provided AEPs and adjusting them to their objectives. Power system IoT AEPs include Siemens' MindSphere, GE Digital's Predix, GridBright, and other solutions. PTC also offers a leading industrial IoT solution, ThingWorx, which can be integrated with other PTC solutions. MathWorks' ThingSpeak is a common AEP for research and prototyping with capabilities similar to industry-grade AEPs such as data collection, storage, analysis, and dashboard creation. Researchers in IoT implementation for power systems favor ThingSpeak because of its integration with MATLAB and Simulink [17]–[20].

2.2.4 The cybersecurity layer

To ensure the safety and reliability of smart grids, it is crucial to implement an additional cybersecurity layer that emphasizes the importance of cybersecurity in IoT infrastructure. This layer should be based on five main goals: authentication, authorization, availability, confidentiality, and integrity. These goals are essential to protect against potential cyber threats that can lead to physical damage, economic losses, and customer data leaks.

Authentication is the process of verifying the identity of a user or device, while authorization is the process of granting access to specific resources or functions. These two goals are crucial in ensuring that only authorized users and devices can access smart grid systems and perform authorized actions. Availability is also critical to ensure that smart grid systems are always accessible and operational, even in the event of a cyber-attack.

Confidentiality and integrity are two other essential goals that are vital to protect the data privacy of smart grid systems. Confidentiality ensures that sensitive data remains private and secure, while integrity ensures that data is accurate and has not been tampered with. Both goals are necessary to protect against potential cyber-attacks that could compromise the data privacy and security of smart grid systems.

The importance of a cybersecurity layer for smart grids cannot be overstated. Smart grids are critical national infrastructures that must be protected against cyber threats. A possible cyber-attack could lead to physical damage, economic losses, and customer data leaks. Therefore, implementing a cybersecurity layer that focuses on authentication, authorization, availability, confidentiality, and integrity is essential to ensure the safety and reliability of smart grid systems.

2.3 IoT Applications within the Power System

The power system is a complex and intricate system of systems that work together to generate, distribute, and manage electricity. With the increasing demand for reliable and efficient energy systems, there is a growing interest in implementing IoT technology to improve the operation and maintenance of power systems.

However, due to the complexity of the power system, it is not straightforward to generalize how IoT can be used in this context. The diverse range of equipment, varying network architectures, and unique operational requirements make it challenging to design IoT solutions that can be easily applied across all power systems. Therefore, a careful and detailed analysis of the

Table 2.4 IoT value chain components in relation to power systems.

IoT value chain component	Description	How it fits within the power system
Device layer	The physical devices or things being monitored or controlled through sensors or actuators.	Devices in a power system can include transformers, switchgear, meters, and other equipment. *Sensors* can be used to monitor parameters such as voltage, current, and temperature, while *actuators* can be used to control equipment, such as circuit breakers.
Communication layer	Facilitates the transmission of data from the device to the cloud-based AEP. It can take various forms, such as cellular networks, Wi-Fi, or satellite communication.	Communication networks can be used to *transmit* data from sensors and other devices in the power system to the AEP. Depending on the application, different types of networks may be used, such as wide-area networks, local-area networks, or wireless networks.
Cloud-based AEP	Provides computational power and storage capabilities for IoT systems. Hosts the necessary algorithms for data analysis and processing and provides a visualization interface for end-users to interpret and utilize the information generated by the system.	The AEP can be used to process and *analyze* data from sensors and other devices in the power system, allowing operators to monitor and optimize system performance. *Visualization* tools can also be used to provide insights into system operations, helping to identify areas for improvement.

specific power system's needs and characteristics is essential to identify suitable IoT applications and ensure their effective implementation.

After discussing the main components of the Internet of Things (IoT) value chain in Section 2.2, Table 2.4 provides a concise yet informative overview of how these components work together within the context of the power system.

However, it is important to delve deeper into each power system part, from generation and transmission to distribution, and explore the application of IoT within each.

2.3.1 IoT applications in the generation stage

For thermal power generation, WSAN can be fitted to the main components of power plants, such as boilers, turbines, generators, condensers, or transformers, to monitor their operation in real time. Although SCADA control systems are already in operation, implementing IoT can capture additional insights that can be utilized to optimize plant efficiency, reduce fuel consumption, and lower emissions [21]–[24].

Implementing IoT in renewable energy generation provides further advantages compared to its thermal counterpart. This is due to the nature of these resources, such as solar or wind generation, which usually consist of multiple connected generation units and are dependent on weather conditions for their power output. This creates the need for central management and optimization of their operation to achieve maximum power output, which is typically achieved using IoT systems [25]–[32].

2.3.2 IoT applications in the transmission stage

Due to the nature of transmission lines spreading over long distances, it has always been equipped with different measurement sensors, such as phasor measurement units, to enable remote monitoring and control and to detect and locate faults. IoT sensors can be installed on transmission lines, transformers, and other equipment to collect real-time data on voltage, current, temperature, and other operating parameters. This data is transmitted to a central monitoring system, which uses machine learning algorithms to analyze the data and detect anomalies or potential faults [33]–[36]. This enables grid operators to take corrective actions before a fault leads to a power outage or other disruptions. Additionally, IoT-enabled predictive maintenance can help grid operators optimize maintenance schedules and reduce maintenance costs. Overall, the implementation of IoT infrastructure within transmission lines can help improve the reliability, efficiency, and safety of the power system.

2.3.3 IoT application in the distribution stage

The implementation of advanced metering infrastructure (AMI) within distribution networks of power systems has brought significant advancements in the monitoring and control of energy distribution. AMI is actually an application of IoT in distribution networks, where smart meters are being deployed to measure and monitor parameters such as voltage, current, and power quality, providing real-time data on the network's performance. This data can be analyzed to identify potential issues and proactively manage the distribution system, resulting in improved efficiency and reliability. Additionally, AMI infrastructure enables the integration of renewable energy sources and storage devices, allowing for the effective management of fluctuating energy supply and demand. Overall, the implementation of AMI in power distribution networks is transforming them from traditionally passive to active intelligent systems that are better equipped to meet the modern type of demand.

2.4 Recent IoT Applications for Modern Power Systems

An IoT-enabled grid, also known as a smart grid, represents the integration of cyber–physical systems with the power grid, enabling higher processing power and greater reliance on information communication technology (ICT). The implementation of a smart grid aims to achieve multiple objectives, including real-time monitoring and control of the grid, rapid fault prevention and system restoration, effective bi-directional power flow management, optimization of economic operation, and dynamic control over both generation and loads.

To better comprehend the key differences between traditional power grids and smart grids, a comparison is provided in Table 2.5.

Recent IoT implementations for modern power system technologies have revolutionized the way energy is produced, stored, and distributed. This is especially important for recent low-carbon technologies being implemented into the grid. The following are some of these applications.

2.4.1 Energy storage systems

IoT devices can monitor various values, including the state of charge, battery health, and energy consumption patterns. By collecting real-time data on energy usage and battery performance, operators can optimize charging and discharging cycles, prevent equipment failures, and ensure reliable power supply. Additionally, energy storage systems can be integrated with renewable energy sources such as solar panels and wind turbines, enabling operators to store excess energy and release it back to the grid when needed.

2.4.2 Wind farms

IoT sensors can monitor wind speeds, turbine performance, and weather conditions, providing valuable insights into wind farm operations. By collecting

Table 2.5 A comparison between traditional and smart grids.

Traditional power grids	Smart grids
Depends on a few centralized high-capacity generation stations feeding all of the connected loads.	Depends on a large number of distributed medium- and small-capacity generators to feed the load.
Mostly dependent on fossil fuels and thermal power plants.	More dependent on clean renewable energy sources.
Due to the nature of generation producing emissions, the generation is located relatively remote from the load center; thus, long transmission lines are required.	The DERs can be located near the loads as they do not cause any environmental issues.
It is designed for a one-way power flow from generation to transmission to distribution.	It is able to coordinate a bi-directional power and information flows, and we can find generation in the low voltage network.
The generation is totally controlled, but the grid operator has no control over the loads.	Renewable generation is intermittent, but its output can be modulated using advanced power electronics devices, and the load can be controlled using demand side management.
Customers are only consumers of the supplied energy.	Customers can be just consumers or prosumers who produce electricity to the electricity market.
It has a limit to maximum connected renewable energy sources. This is referred to as the hosting capacity. It also has a lot of conservations when it comes to connecting electric vehicles (EVs) to the grid.	It is designed with renewable energy sources and electric vehicles in mind. As to accommodate renewables stochastic nature and to manage EVs charging and discharging while maintaining a stable grid operation.
Depends on measuring consumption through traditional electricity meters.	Depends on a network of AMI that sends different consumption information about each consumer, even each connected appliance on a basis of a relatively small fixed time step.
Dispatches generation according to a given load demand predicted from load profiles, and hot reserve is maintained for unpredicted load changes.	Controls both generation and load sides in order to achieve optimal operation of the whole system considering technical and economic constraints.
Electricity retail price is determined by a contractual value.	Retail price is varying in accordance to the cost of generation, with incentives for customer for higher energy efficiency.
Relies on one big network or grid that connects the whole city or country.	Can be made up of a group of microgrids, connected or isolated from the common grid.

real-time data, operators can optimize turbine performance, predict maintenance needs, and prevent equipment failures. Additionally, IoT solutions can enable wind farms to respond quickly to changes in wind speeds and direction, ensuring reliable power generation.

2.4.3 Electric vehicle fleets

IoT-enabled charging stations can monitor the state of charge and charging status of electric vehicles, enabling fleet operators to optimize charging schedules and prevent battery degradation. Vehicle-to-grid technology allows EVs to store and supply energy to the grid during peak demand periods, providing a valuable source of energy storage for the grid. Additionally, IoT devices can monitor vehicle performance and maintenance needs, enabling fleet operators to optimize vehicle operations and reduce downtime.

2.4.4 Advanced metering infrastructure

Implementing smart meters in the distribution network transforms network operations from a more passive "fit and forget" approach, where generating units are set according to a predicted load curve, into an active interaction that takes near real-time action to change demand on a half-hourly basis. This is necessary due to changes in demand brought about by embedded generation, energy storage, and electric vehicle charging stations. Smart meters and the underlying ICT infrastructure that comprises the AMI are perfect examples of the IoT value chain.

The IoT value chain begins with the smart meter, which creates a home area network within each customer's house. This is essentially the WSAN for this application, where smart electricity and gas meters are connected together with an in-home display and other smart connected appliances. These typically use a ZigBee network [37]–[40]. Measured data is then sent using a communication transport, commonly a cellular network, to the distribution area network for operational monitoring and control within the distribution network. Data from each distribution area network is collected centrally and monitored by the grid operator to maintain balancing services between generation and demand.

Overall, IoT solutions are transforming modern power system technologies by enabling real-time monitoring and control, optimizing performance, and improving reliability and sustainability. This is not only limited to the three examples previously mentioned; however, IoT technology keeps finding its way into the smallest parts of the power systems.

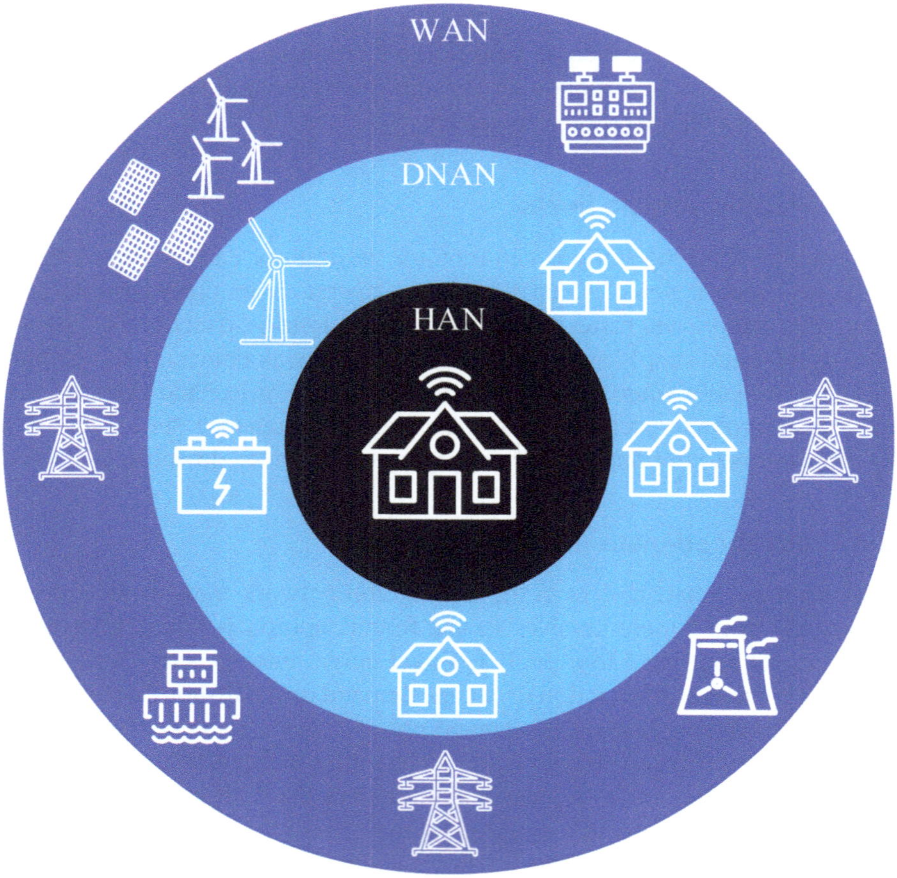

2.5 Conclusion

This chapter explored the infrastructure of Internet of Things (IoT) technologies in modern power systems, beginning with establishing a comprehensive understanding of the various components comprising the IoT value chain. The device layer, communication layer, cloud layer, and cybersecurity layer are the essential components of the IoT value chain. The device layer is responsible for determining the specific application of the IoT system, and hardware requirements include a sensor, actuator, micro-controller unit, memory, and gateway. The communication layer is necessary for facilitating signal transport between IoT devices and the cloud, while the 5G-IoT represents a significant potential for IoT implementation in sub-second applications of control and protection. The cloud layer serves as a foundation for the application enablement platform, providing the essential infrastructure for device connectivity and data management. Finally, the cybersecurity layer is crucial in protecting IoT infrastructure from potential cyber threats; thus, the cybersecurity layer is as critical as the underlying IoT application. Careful consideration of the specific requirements of each component can lead to a system that meets the needs of end-users while providing the necessary insights and analytics for informed decision-making.

From generation to distribution, IoT has revolutionized the way energy is produced, stored, and distributed, enabling operators to optimize performance, reduce costs, and improve reliability. By implementing IoT sensors and devices, operators can collect real-time data on energy usage, battery performance, wind speeds, vehicle charging, and more, enabling them to make informed decisions and take proactive actions. Furthermore, IoT-enabled predictive maintenance can help operators optimize maintenance schedules and reduce downtime, and energy storage systems can be integrated with renewable energy sources to ensure a reliable power supply.

Implementing AMI within distribution networks transforms them from traditionally passive to active intelligent systems that are better equipped to meet modern demand. In conclusion, the IoT infrastructure is crucial in modern power systems, enabling operators to monitor and control operations in real time, improve efficiency and reliability, and reduce costs. As technology continues to advance, we can expect to see further IoT innovations in the power sector, providing new opportunities for optimization and improved performance.

References

[1] R. Wenzlhuemer, 'The Development of Telegraphy, 1870–1900: A European Perspective on a World History Challenge', *Hist. Compass*, vol. 5, no. 5, pp. 1720–1742, 2007, doi: 10.1111/j.1478-0542.2007.00461.x.

[2] W. Aspray, 'Inventing the internet', *IEEE Spectr.*, vol. 34, no. 2, pp. 12–12, Feb. 1997, doi: 10.1109/MSPEC.1997.570802.

[3] L. Columbus, 'Roundup Of Internet Of Things Forecasts And Market Estimates, 2016', *Forbes*. https://www.forbes.com/sites/louiscolumbus/2016/11/27/roundup-of-internet-of-things-forecasts-and-market-estimates-2016/ (accessed Mar. 13, 2023).

[4] C. Cecati, G. Mokryani, A. Piccolo, and P. Siano, 'An overview on the smart grid concept', in *IECON 2010 - 36th Annual Conference on IEEE Industrial Electronics Society*, Nov. 2010, pp. 3322–3327. doi: 10.1109/IECON.2010.5675310.

[5] X. Yu and Y. Xue, 'Smart Grids: A Cyber–Physical Systems Perspective', *Proc. IEEE*, vol. 104, no. 5, pp. 1058–1070, May 2016, doi: 10.1109/JPROC.2015.2503119.

[6] 'Grid Modernization and the Smart Grid', *Energy.gov*. https://www.energy.gov/oe/grid-modernization-and-smart-grid (accessed Mar. 13, 2023).

[7] J. A. Momoh, 'Smart grid design for efficient and flexible power networks operation and control', in *2009 IEEE/PES Power Systems Conference and Exposition*, Mar. 2009, pp. 1–8. doi: 10.1109/PSCE.2009.4840074.

[8] M. Badawi *et al.*, 'Reliable Estimation for Health Index of Transformer Oil Based on Novel Combined Predictive Maintenance Techniques', *IEEE Access*, vol. 10, pp. 25954–25972, 2022, doi: 10.1109/ACCESS.2022.3156102.

[9] R. J. Tom, S. Sankaranarayanan, and J. J. P. C. Rodrigues, 'Smart Energy Management and Demand Reduction by Consumers and Utilities in an IoT-Fog-Based Power Distribution System', *IEEE Internet Things J.*, vol. 6, no. 5, pp. 7386–7394, Oct. 2019, doi: 10.1109/JIOT.2019.2894326.

[10] N. Nimalsiri, 'Electric Vehicle (EV)-Assisted Demand-Side Management in Smart Grid', Ph.D., The Australian National University (Australia), Ann Arbor, 2021.

[11] K. T., C. R. S., J. D. N. J., and C. K., 'Design of IoT based smart compact energy meter for monitoring and controlling the usage of energy and power quality issues with demand side management for a commercial building', *Sustain. Energy Grids Netw.*, vol. 26, p. 100454, Jun. 2021, doi: 10.1016/j.segan.2021.100454.

[12] J. P. Lahti, P. Helo, A. Shamsuzzoha, and K. Phusavat, 'IoT in electricity supply chain: Review and evaluation', in *2017 15th International Conference on ICT and Knowledge Engineering (ICT&KE)*, Nov. 2017, pp. 1–6. doi: 10.1109/ICTKE.2017.8259615.

[13] O. Kazeem, O. Akintade, and L. Kehinde, 'Comparative Study of Communication Interfaces for Sensors and Actuators in the Cloud of Internet of Things', *Int. J. Internet Things*, vol. 6, pp. 9–13, Jun. 2017, doi: 10.5923/j.ijit.20170601.02.

[14] J.-S. Lee, Y.-W. Su, and C.-C. Shen, 'A Comparative Study of Wireless Protocols: Bluetooth, UWB, ZigBee, and Wi-Fi', in *IECON 2007 - 33rd Annual Conference of the IEEE Industrial Electronics Society*, Nov. 2007, pp. 46–51. doi: 10.1109/IECON.2007.4460126.

[15] M. Faheem *et al.*, 'Smart grid communication and information technologies in the perspective of Industry 4.0: Opportunities and challenges', *Comput. Sci. Rev.*, vol. 30, pp. 1–30, Nov. 2018, doi: 10.1016/j.cosrev.2018.08.001.

[16] F. E. Abrahamsen, Y. Ai, and M. Cheffena, 'Communication Technologies for Smart Grid: A Comprehensive Survey', *Sensors*, vol. 21, no. 23, Art. no. 23, Jan. 2021, doi: 10.3390/s21238087.

[17] S. Li, L. Da Xu, and S. Zhao, '5G Internet of Things: A survey', *J. Ind. Inf. Integr.*, vol. 10, pp. 1–9, Jun. 2018, doi: 10.1016/j.jii.2018.01.005.

[18] 'Siemens | MindSphere'. https://siemens.mindsphere.io/en (accessed Jan. 30, 2023).

[19] 'Home - Gridbright'. https://www.gridbright.com/ (accessed Jan. 30, 2023).

[20] 'ThingWorx: Industrial IoT Software | IIoT Platform | PTC'. https://www.ptc.com/en/products/thingworx (accessed Jan. 30, 2023).

[21] G. Bedi, G. K. Venayagamoorthy, R. Singh, R. R. Brooks, and K.-C. Wang, 'Review of Internet of Things (IoT) in Electric Power and Energy Systems', *IEEE Internet Things J.*, vol. 5, no. 2, pp. 847–870, Apr. 2018, doi: 10.1109/JIOT.2018.2802704.

[22] J. Prihantoro, H. Taufiqurrohman, Suyanto, Cahyadi, A. Larasati, and A. Nurrohim, 'Implementation SCADA Web Client For Monitoring Geothermal Power Plant 3 MW Kamojang - Indonesia', in *2022 2nd International Conference on Electronic and Electrical Engineering and Intelligent System (ICE3IS)*, Nov. 2022, pp. 270–275. doi: 10.1109/ICE3IS56585.2022.10010311.

[23] P. Sharmila, C. Nayanatara, P. Shanmugapriya, A. Aravind, and V. M. Lavanya, 'Power Plant Automation Using SCADA in a Gas Insulated Substation', in *2022 International Conference on Communication, Computing and Internet of Things (IC3IoT)*, Mar. 2022, pp. 01–05. doi: 10.1109/IC3IOT53935.2022.9767890.

[24] 'IoT Solutions for Electric Power Industry | softengi.com'. https://softengi.com/blog/iot-solutions-for-the-electric-power-industry/ (accessed Mar. 13, 2023).

[25] A. Malik, A. Haque, and V. S. B. Kurukuru, 'IoT-Based Monitoring and Management for Photovoltaic System', in *Fault Analysis and its Impact on Grid-connected Photovoltaic Systems Performance*, John Wiley & Sons, Ltd, 2022, pp. 291–318. doi: 10.1002/9781119873785.ch9.

[26] I. Abdullahi, S. Perinpanayagam, and I. Hamidu, 'A Fog Computing Based Approach Towards Improving Asset Management and Performance of Wind Turbine Plants Using Digital Twins', in *2022 27th International Conference on Automation and Computing (ICAC)*, Sep. 2022, pp. 1–5. doi: 10.1109/ICAC55051.2022.9911170.

[27] A. Goudarzi, F. Ghayoor, M. Waseem, S. Fahad, and I. Traore, 'A Survey on IoT-Enabled Smart Grids: Emerging, Applications, Challenges, and

Outlook', *Energies*, vol. 15, no. 19, Art. no. 19, Jan. 2022, doi: 10.3390/en15196984.

[28] S. Voutsinas, D. Karolidis, I. Voyiatzis, and M. Samarakou, 'Development of an IoT power management system for photovoltaic power plants', in *2022 11th International Conference on Modern Circuits and Systems Technologies (MOCAST)*, Jun. 2022, pp. 1–5. doi: 10.1109/MOCAST54814.2022.9837652.

[29] M. J. Ali, A. Mondal, and P. Dutta, 'Intelligent Monitoring and Control of Wind Turbine Prototype Using Internet of Things (IoT)', in *2022 IEEE International IOT, Electronics and Mechatronics Conference (IEMTRONICS)*, Jun. 2022, pp. 1–6. doi: 10.1109/IEMTRONICS55184.2022.9795845.

[30] M. O. Qays *et al.*, 'Monitoring of renewable energy systems by IoT-aided SCADA system', *Energy Sci. Eng.*, vol. 10, no. 6, pp. 1874–1885, 2022, doi: 10.1002/ese3.1130.

[31] M. Habib, A. Gram, A. Harrag, and Q. Wang, 'Optimized management of reactive power reserves of transmission grid-connected photovoltaic plants driven by an IoT solution', *Int. J. Electr. Power Energy Syst.*, vol. 143, p. 108455, Dec. 2022, doi: 10.1016/j.ijepes.2022.108455.

[32] J. Hossain, N. Algeelani, A. Al-Masoodi, and A. Kadir, 'Solar-wind power generation system for street lighting using internet of things', *Indones. J. Electr. Eng. Comput. Sci.*, vol. 26, p. 639, May 2022, doi: 10.11591/ijeecs.v26.i2.pp639-647.

[33] N. Dhanalakshmi, C. Mamindlapalli, D. R. Sunkari, and R. R. Salguti, 'Design of IoT-Based Transmission Line Fault Monitoring System', in *Advances in Signal Processing and Communication Engineering*, Singapore, 2022, pp. 373–385. doi: 10.1007/978-981-19-5550-1_35.

[34] N. Sivabalan, S. Saravanan, R. Thirukkumaran, and R. Jaya, 'Novel study of transmission line monitoring using IoT', *AIP Conf. Proc.*, vol. 2519, no. 1, p. 030043, Oct. 2022, doi: 10.1063/5.0110665.

[35] X. Kong, Y. Xu, Z. Jiao, D. Dong, X. Yuan, and S. Li, 'Fault Location Technology for Power System Based on Information About the Power Internet of Things', *IEEE Trans. Ind. Inform.*, vol. 16, no. 10, pp. 6682–6692, Oct. 2020, doi: 10.1109/TII.2019.2960440.

[36] P.-Y. Kong and Y. Song, 'Artificial-Neural-Network-Assisted Sensor Clustering for Robust Communication Network in IoT-Based Electricity Transmission Line Monitoring', *IEEE Internet Things J.*, vol. 9, no. 17, pp. 16701–16713, Sep. 2022, doi: 10.1109/JIOT.2022.3150888.

[37] R. K. Pillai, R. Bhatnagar, and H. Thukral, 'AMI rollout strategy and cost-benefit analysis for India', in *2016 First International Conference*

on Sustainable Green Buildings and Communities (SGBC), Dec. 2016, pp. 1–6. doi: 10.1109/SGBC.2016.7936083.

[38] T.-S. Choi, K.-R. Ko, S.-C. Park, Y.-S. Jang, Y.-T. Yoon, and S.-K. Im, 'Analysis of energy savings using smart metering system and IHD (in-home display)', in *2009 Transmission & Distribution Conference & Exposition: Asia and Pacific*, Oct. 2009, pp. 1–4. doi: 10.1109/TD-ASIA.2009.5356956.

[39] N. Matanov and A. Zahov, 'Remote Electricity Metering Systems', in *2019 11th Electrical Engineering Faculty Conference (BulEF)*, Sep. 2019, pp. 1–6. doi: 10.1109/BulEF48056.2019.9030800.

[40] C. Y. Chang, *The Evidence Base Benefit of the AMI to the Users and Power Network Efficiency -The Case of the UK*. 2022.

3

Cyber–Physical Systems

K. Selvasundar[1] and T. Deepa[2]

[1]School of Electrical Engineering, Vellore Institute of Technology, India
[2]School of Electrical Engineering, Vellore Institute of Technology, India

Emails: selvasundar.k2020@vitstudent.ac.in; deepa.t@vit.ac.in

Abstract

A cyber–physical system (CPS) is an innovative variant of digitalization that has attracted interest from governments, research, and industries. As cyber–physical power systems (CPPSs) have evolved, conventional power networks, which include physical hardware as a core component, have become more interconnected with communication and information technology. A CPPS is a heterogeneous network that is strongly coupled to cyber systems and allows bidirectional power and information flows, making microgrids possible. Despite the fact that digital technologies are more efficient and dependable, the electrical grid is susceptible to cybersecurity threats and requires a synergy between cyber and physical systems. Modeling approaches and a thorough evaluation of the interaction between cyber and physical systems are essential for the analysis and resolution of CPPS glitches. Modeling, simulation, and analytic approaches that distinguish the physical and cyber realms are incompatible with CPPSs. Thus, a groundwork for integrating physical and cyber systems needs to be developed to examine the practical implications of combining them. The chapter discusses a variety of modeling, simulation, and analytic tools together with various forms of cyber-attacks and cybersecurity actions for current CPPSs. This study provides an overview of the various detection and control techniques for detecting and mitigating cyber-attacks on power systems.

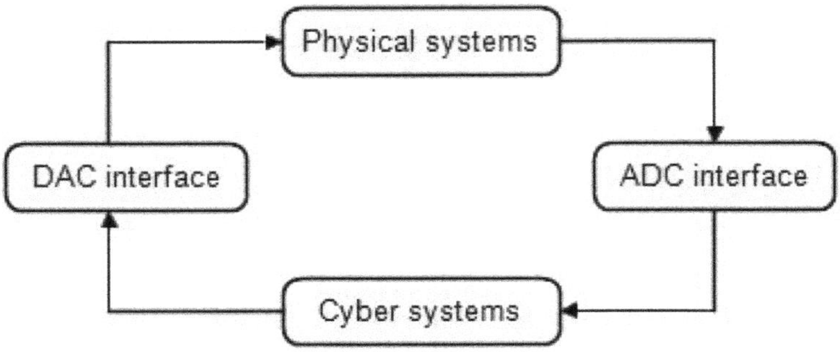

Figure 3.1 Block diagram of cyber–physical systems.

3.1 Introduction

Experts have been developing tools and strategies to enhance the conduct of observing and governing physical power systems in recent years. Engineers in the fields of computer science and electronics work together to improve the computing and communication systems simultaneously. As a result, ubiquitous computing is being developed. Every gadget and technological device in our daily lives is interconnected to low-cost processing and communication networks. This will have an immense impact on energy systems [1]. The CPS is a new digital technology created by combining physical and cyber systems. As part of its expansion, CPS is now paying more attention to all areas, such as agriculture, energy, medicine, the oil and gas industry, and transportation. It is a multi-dimensional heterogeneous system that uses cyber components to provide stability, resilience, efficiency, and dependability in physical systems.

A CPS gathers information from the physical systems and transmits it to cyber systems via an analog-to-digital converter (ADC) interface, as shown in Figure 3.1. Linking physical and cyber systems warrants an effective and consistent operation of power systems [2].

A CPPS gets established when a physical electrical power system and a cyber system are integrated [3], [4]. The CPPS encompasses all aspects of electric power systems, including production, distribution, and consumption. A CPPS is an internet and physical power system integration and coordination system. These are decentralized networks that exhibit an unusual behavior. The CPS comprises control and integrated devices that regularly monitor the physical power system, as shown in Figure 3.1. CPPS is made up of interconnected components with physical input and output. To generate new scientific, technological, and creative products, computing and networking are coupled with traditional power systems. A system that integrates connectivity,

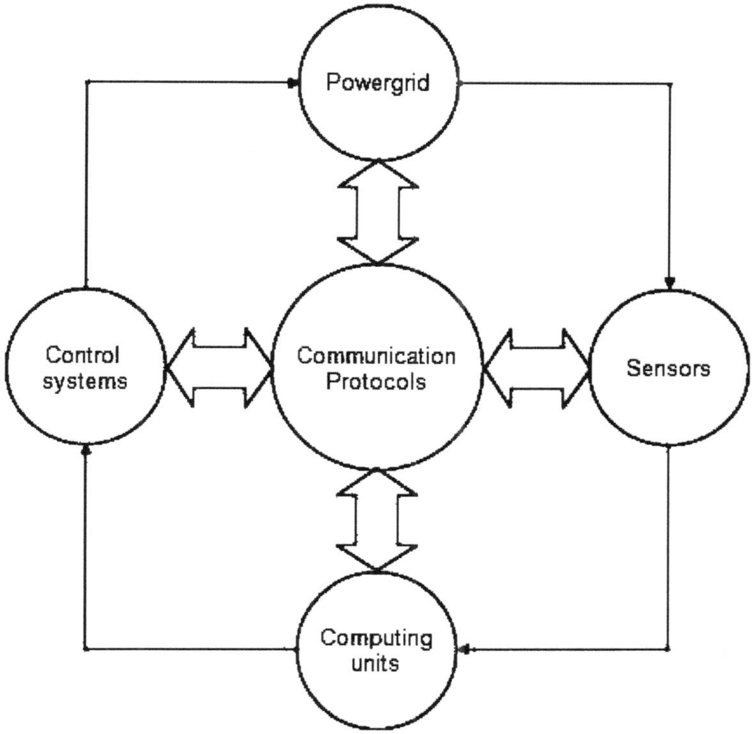

Figure 3.2 Structure of cyber–physical power systems.

processing, and manipulation is referred to as "cyber." Traditional power systems operate swiftly and are governed and managed by physical principles.

The CPPS interconnection protocol is used to connect heterogeneous systems using a common communication channel. CPS that requires complete instruction and performance assurance for cyber–physical interactions, such as CPPSs, should use this protocol. This protocol's principal goal is to provide CPPSs with heterogeneity on three levels: function compatibility, integration, and performance assurance. The CPS interconnection protocol was eventually developed using the transport protocol services. Individual comprehension of the power system's physical and computational components is insufficient. Instead, we must appreciate their interaction, as shown in Figure 3.2.

A power system controller collects data from its main components and translates it into a control signal, which is then transmitted back to the fundamental components for optimal power grid operation. A subsequent level of interaction exists between the communication infrastructure and power system control. All of the subsystem functions are coordinated by the communication infrastructure in CPPSs. Data loss, faulty data, significant delays, and other communication issues impede the CPPS performance [5]. A third level of interaction exists in between communication networks and the cyberspace system. Master and slave networks, master client, communicative server, two-way communication structure, computer literacy stations, sophisticated control software applications, computer hackers defense and security mechanisms, and so on are some of the components of cyber systems. The major goal of the cyber system is to do complicated grid tasks such as load flow analysis, state prediction, parameter management, voltage regulation, oscillatory supervision, broad tracking and management, operational processes, validation of the model, and transient stability. Increased power grid network growth to meet load demand makes expanding the cyber system easier. Despite the fact that the cyber system assures the power grid's effective, secure, and reliable operation, cyber system failure has resulted in power shortages in the past.

The fundamental flaw of CPPS relies on cyber-attacks and security issues. A CPPS is a massive hybrid interconnected power transmission system capable of being hacked. Due to flaws in cybersecurity features, cyber system components are extremely exposed to cyber threats and attacks via cyber connections. Since the cyber-attack does not explicitly interrupt the actual power supply, it has the similar effect as physical harm and produces system instability when combined with a physical attack. As a result, a CPPS must evaluate all types of cyber threats as well as preventative actions.

Researchers from all over the world have looked into CPPS from a variety of angles. The substantial interdependency between cyber and physical systems is a key feature of CPPS. Using a model-based technique, we evaluated the effects of several cyber scenarios on a physical system. To defend CPPS's advanced power grid control networks from cyber-attacks, risk and vulnerability assessments are required. It is the CPPS's comprehensive cybersecurity component that uses advanced information automated reasoning algorithms to differentiate itself from traditional information security. It can tell the difference between lawful and malicious online behavior. Designing a wide-area damping controller for mitigating inter-area fluctuations in wide-range CPPS is gaining popularity in the event of a cyber-attack on the physical power system. Before the CPPS turns black, the cyber–physical attack resistive wide-area control technology tries to strengthen the system's reliability. Because of its high dimensionality and sophisticated connectivity structure, it is designed to adapt to the present CPPS's continual expansion while also taking into account cyber contingencies on the practical power supply.

3.2 Modeling of CPPS

The complex interaction between physical and cyber systems at diverse times, places, and sizes is one of the most significant aspects of CPPS modeling. A dynamic physical system consists of a generation, transformers, transmission system, user, and other elements directly associated with energy flow. A static cyber system is made up of cyber components connected by a telecommunication network that permits data to flow constantly [6]. The intimate link between the physical and cyber systems in the CPPS acts as a crucial failure point because both systems are in distinct topologies. When one system in a wide-range CPPS fails, the entire system fails catastrophically [7].

Both physical and cyber systems have their own set of risks. The assimilation of renewable power into the physical process, which is random in nature, has an influence on the power flow in the system's stable operating condition [8]. Cyber-attacks against the controlling, processing, and communication functions of cyber systems disrupt the data flow. These variables are unknown, posing a greater risk to the electrical system's safe and stable operation. CPPS modeling is complicated by the collaborative elements of physical and cyber systems.

As a result, qualitative and quantifiable evaluations of physical and cyber systems are required. The modeling background for a critical understanding of complexity and interdependence in CPPS is also essential. This will help to avert catastrophic cascade failure scenarios in a networked CPPS. The modeling of the CPPS can be divided into three categories:

i. CPPS interconnection modeling

ii. CPPS interaction modeling

iii. CPPS interdependent modeling

3.2.1 CPPS interconnection modeling

The CPPS in this model represents the connectivity of physical and cyber systems, along with the necessity to connect them. The power system's physical components must be monitored and managed in the physical system [9]. A cyber system comprises computation, control, and communication algorithms. ADC, digital-to-analog converter, and digital networks are among the systems required to connect physical and cyber systems. Differential equations for regular interaction of physical processes are combined with difference equations for periodical performance of converters, digitized networks, and cyber systems in hybrid dynamical system theory [10]. It is categorized as follows.

3.2.1.1 Physical components modeling

It is a continuous-time system in which the variables, or states, are parameterized by a difference equation. Limiting the status and input values to the physical system is necessary in some instances.

3.2.1.2 Cyber component modeling

Cyber components are responsible for running algorithms, performing computations, and sending data through digital networks. The cybernetic component state variables are discrete values that are changed at discrete occurrences rather than on a continuous basis.

3.2.1.3 Finite State Machine (FSM)

It is a scientific concept that depicts the interaction among a system's input and output. In different applications, it is used to define the controlled program execution and modeling of sequential circuits. The input and output of the FSM are changed for each value of the input. The FSM's states, inputs, and outputs are all discrete sets. They are updated when its incoming data generates discrete transitions.

3.2.2 CPPS interaction modeling

Individual cyber and physical components' models are combined via interfaces to make a comprehensive mathematical model of the CPS. The CPS feedback connectivity modeling is shown in Figure 3.3.

In the paradigm, the cyber and physical systems have diverse dynamics, with the cyber system having discontinuous attributes and the physical system having continuous dynamics [11]. Cyber and physical systems are linked by interfaces, which convert signals appropriately. To connect the cyber and physical systems, a mathematical model of interfaces is employed, and the cyber system, physical system, and interfaces are all organized to create the CPS comprehensive model.

The various forms of CPPS interaction models are discussed in this section. The CPPS interaction model is divided into five kinds based on the literature evaluation, as shown in Figure 3.4.

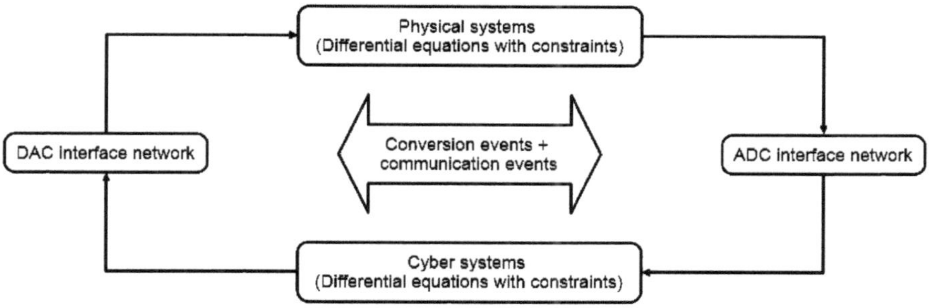

Figure 3.3 CPPS interaction modeling.

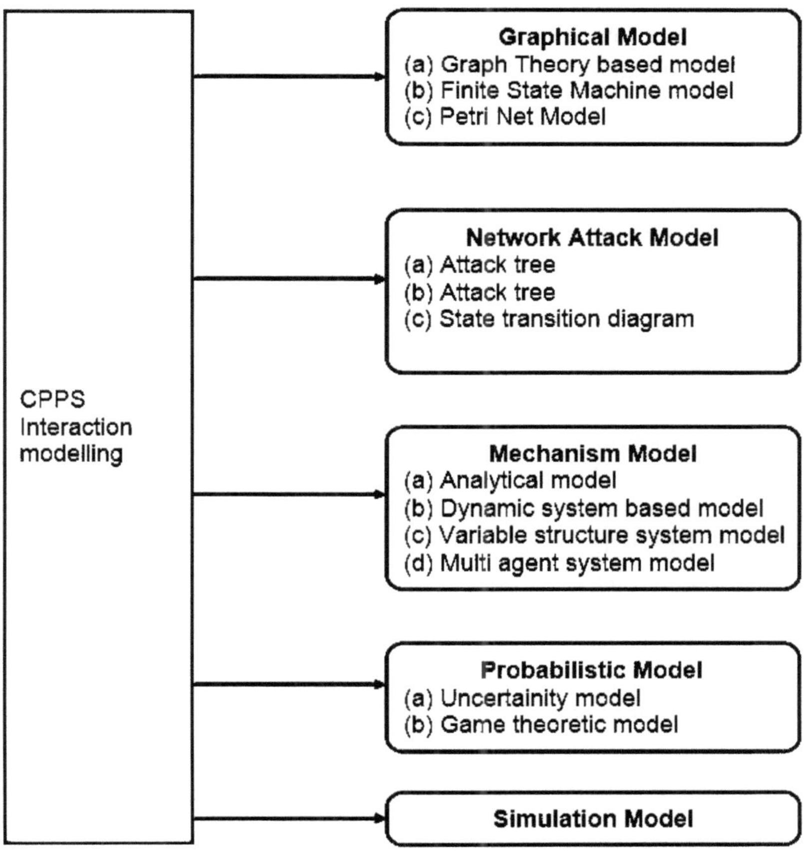

Figure 3.4 Structure of CPPS interaction modeling.

There are five different kinds of models: graphical model, network attack model, mechanism model, probabilistic model, and simulation model.

3.2.2.1 Graphical modeling

The visualization-based interaction between physical and cyber systems is shown in a graphical model. It aids in the construction of the power grid's structure and the analysis of the grid's operation in the face of several threats [12].

3.2.2.1.1 Graph-theory-based model

In CPPS, transmission lines connect generators, breakers, safety relays, and loads, whereas communication networks connect cyber system components. Each element of the physical phenomenon is anticipated to be integrated with the cyber node for monitoring and controlling CPPS [13]. It uses routers and switches to provide constituent information to the remote-control center. When the signal is obtained in the central controller, it is evaluated, and a command signal is produced, which is subsequently directed to control devices such as flexible AC transmission devices.

Since physical and cyber systems interact one-to-one, one system's failure could have an impact on the others and vice versa. The inherent links between physical and cyber systems in CPPS could be investigated using a graph-theory-based method.

3.2.2.1.2 Finite state machine model

A state machine, often known as an FSM, is a computational mathematical concept. The FSM is utilized in a range of applications and conducts a planned set of actions in response to a series of deeds. At any given time, it can be in any of a finite number of states. When it goes from one state to another in response to inputs, it is said to be in state transition. Deterministic and non-deterministic FSMs are the two forms of FSM [14]. For cyber and physical systems, the FSM produces the state chart diagram, which depicts the system's dynamic behavior as it moves through phases during its life cycle.

3.2.2.1.3 Petri net model

Petri nets [15] are a numerical modeling framework to express state transitions and changes in distributed and parallel systems. It is a discrete-event dynamic system that depicts the interconnections in a large-scale system between occurrences, situations, and regulating actions. The Petri net model, according to CPPS [16], is a useful linguistic tool for studying interactions among the continuous nature of the physical system and the discrete character

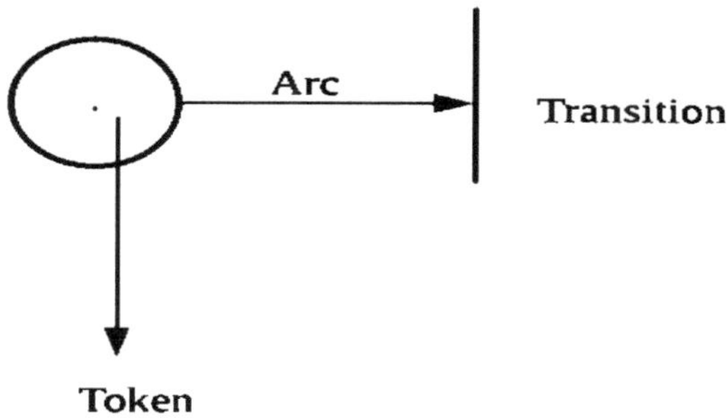

Figure 3.5 Petri net model.

of the cyber system. Petri nets are a graphical modeling approach for regulating asynchronous, concurrent, or distributed CPPS events.

As shown in Figure 3.5, the Petri net model is made up of four basic components: place, transition, arc, and token. Transitions are represented graphically as a bar, arcs are oriented line segments, and tokens are presented graphically as dots. Locations in CPPSs indicate components and their states. In CPPSs, transitions are used to express discrete events that can lead to various states. There are input and output functions. The arcs show how the locations and transitions are connected. Finally, the tokens are used to identify the Petri net's dynamic state, also known as the net's marking [17].

3.2.2.2 Network attack model

Over the previous decade, the CPPS has adopted more complicated technologies to refine the system's operating competence and dependability [18]. A cyber-attack on CPPS is defined as a structured action of tracking the communications system or controlling instructions without authorization and leveraging the program's vulnerabilities to disrupt or limit its function. Cyber-attacks cause system shutdowns by degrading the smart grid's performance. Due to the intricate set of mechanisms between the physical and cyber systems, the breakdown of the cyber network has a significant effect on the physical system. The CPPS behavior may be altered as a result of network assaults, putting the system in an unsafe state and causing harm [19]. As a result, it is essential to evaluate the various forms of cyber-attack models for CPPS utilizing modeling and simulation approaches in order to assess the consequences of cyber-attacks and their implications on CPPS functions such as security, dependability, and profitability. The cyber-attack model aids in identifying and evaluating CPPS's resilience to cyber-attacks. This model is used by power system engineers to:

a. detect an issue at the element and subsystems level and predict a CPPS cyber-attack;

b. improve spatial awareness to safeguard CPPS from upcoming cyber-attacks;

c. assess an electric grid cyberspace domain's cybersecurity;

d. plan and build a more robust CPPS.

The following sections cover the various types of network attack modeling in CPPSs.

3.2.2.2.1 Attack tree

An attack tree is a graphical depiction of all conceivable cyber-attack paths in the power system. It contributes to a new perspective on cyber network infiltration and architecturally and intuitively depicts the cyber-attack process [21]. To assess the liability and risk of critical components of the CPPS, apply the attack tree method [20], [22], [23]. To create a cyber-physical threat model for power system emergencies attack tree method is used. On the other hand, it is only suitable for modeling a certain type of attack and cannot be used to simulate simultaneous or coordinated attacks on several components.

The attack tree for smart grid applications is depicted in Figure 3.6 [24]. Level 1 denotes continuous power transmission to the consumer with no interruptions. Level 2 consequences include changes in the exciter and prime mover reference values to anomalous values, which can lead to a power grid outage. Cyber-attacks on CPPSs that cause physical injury are classified as Level 3. The attacker obtains control of the exciter and prime mover by hacking the supervisory control and data acquisition system and remote terminal unit, causing damage to the power generation system. Finally, the level 4 attack method was used to carry out the attack.

3.2.2.2.2 Attack graph

The attack graph depicts an attacker's behavior and examines the various methods the attacker can employ to exploit system flaws and achieve the desired state. It is an assortment of attack situations in a computer network that each represents a set of events captured by an intruder to access the system with the objective of interrupting customer support, trying to gain access to a sensitive information database or to the main host, among other things. This model uses network topology information to calculate the likelihood of a defect being detected by an attacker and used to carry out incursion and penetration. The attack graphs are used by the system operator to determine the best security solutions to defend their systems. The attack graph model uses an automatic generation approach to find network faults as the network

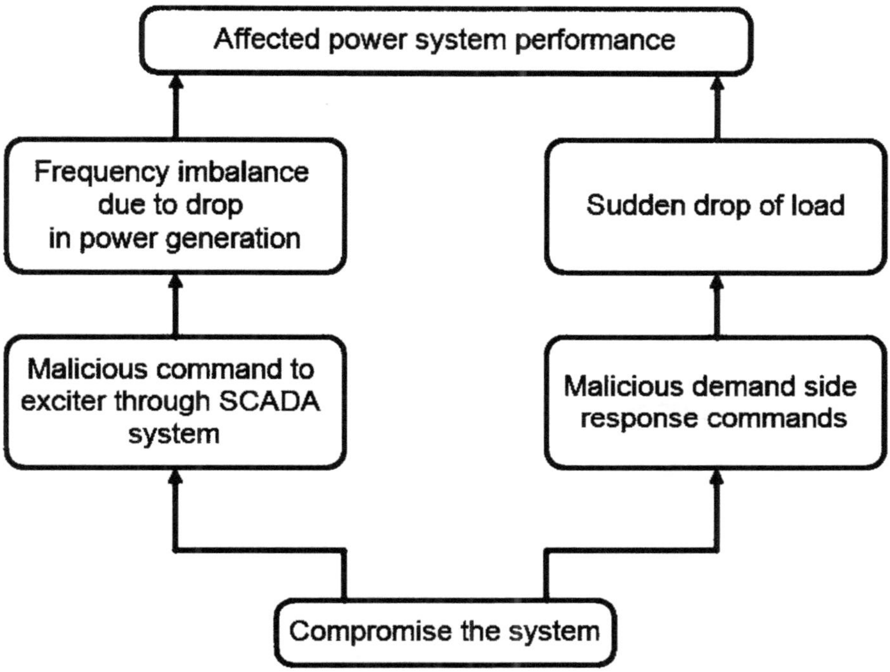

Figure 3.6 Attack tree.

grows larger to replicate massive complicated network attacks. The attack graph model [25] is used to assess the security of the electrical power systems control unit. The automatic generation approach is supplemented with an attack graph model [26] to investigate the influence of rapid failures in the CPPS.

3.2.2.2.3 State transition diagram

The behavior of an assault is characterized in this approach as a Markov decision process, comparable to attack graph methods. Because the states of the Markov process are hidden and cannot directly view the model's state, yet the model's output is dependent on the current state. For evaluating system vulnerabilities, the Markov model forecasts the attack behavior by taking into account the likelihood of the system's state transition under various assault behaviors [27], [28]. The CPPS detection standards are met by this model, which covers all forms of assaults.

Figure 3.7 illustrates an example of a Markov model for assault forecasting in the form of a graph. It depicts four attack states, ranging from

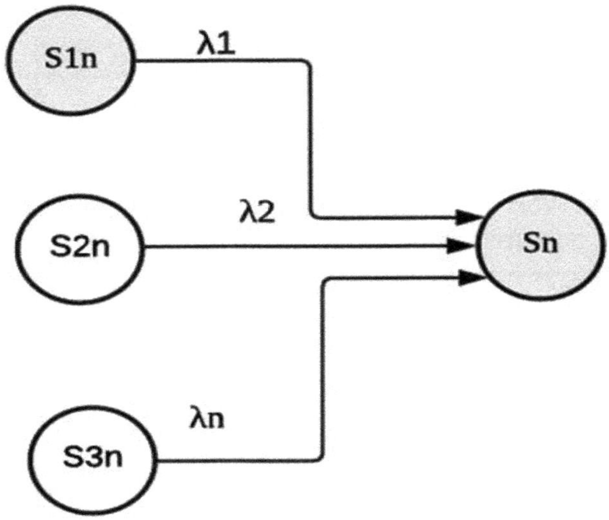

Figure 3.7 Attack graph model.

normal to successful compromise (cyber-attack). Authentication, host and service probes, manipulation, and more are all part of the attack chain. An attacker's next state can be estimated and establish the most likely route from the current state using attack sequences.

3.2.2.3 Mechanism model

Mechanism models in CPPS can be used to investigate interactions between cyber and physical systems, coupling of continuous and discrete event behavior, and the blending of energy and information flow.

3.2.2.3.1 Analytical model

For dependability reasons, both electrical and cyber components in the CPPS are energized by the independent power supply. The consequences of cyber network dependability have an impact on how the cyber and physical systems interact in CPPSs. If an attacker uses synchronized measurement data, oscillatory monitoring outcomes, electrical regulatory prices, and state estimate reports, for example, an attacker can control supervisory control components and tamper with key information about the position of the electrical grid. Failure in cyber network is often seen as an effectuation associated with some operation of the power system in the analytical model framework, and it is utilized to construct a specific power system.

3.2.2.3.2 Dynamic-system-based model

A differential equation with energy transfer represents the physical system in CPPS stability research, while a differential equation with information flow represents the cyber system. Through the stimulation of generator states, the disruption effect of a cyber system on a physical system is modeled. Cyber attackers devised a covert cyber-attack vector to impact the synchronous generator of cyber-controlled distributed energy resources (DERs) in order to preserve the physical uncertainty of the smart grid in [29].

In order to model the CPPS as a closed-loop dynamic system with data and energy flow interconnections, driving devices like exciters, balancers, prime movers, synchronous generators, and HVDC and FACTS devices can be dynamically modeled.

3.2.2.3.3 Variable structure system model

The topology of the CPPS is determined by the state of the circuit breaker switches. The design of the power grid is always evolving, and the dynamics of the power grid depend on the value of the switching signals, which are targeted by an attacker. The flexible structure theory is utilized to calculate the CPPS susceptibility when switching assault signals reconfigure the grid. It demonstrates how CPPS's physical and cyber components interact as well as how the switching vulnerability quickly affects CPPS operations.

3.2.2.3.4 Multiagent system model

The CPPS distribution features become increasingly obvious as the number of DER integrations into CPPS grows, power system operators exchange data more often, and the control scheme becomes more significant. Controlling distinct types of DERs in the CPPS via old unified manner of control is extremely complex and inefficient. The unified control scheme needs the CPPS's thorough mathematical model. With the constant development of sophisticated CPPS, the architecture of a single unified controller for numerous DERs has the potential to fail, as the system has no alternate sources of control, rendering it unstable. Implementing a centralized controller is quite expensive [16]. This prompts the development of a multiagent-based CPPS control system that handles the cooperative optimization challenge of integrating many DERs into CPPS.

3.2.2.4 Probabilistic model

Game-theoretic models and uncertainty models are two types of probabilistic models. All CPPS event types are probabilistic. It is tough to adequately characterize these physical and cyber occurrences. The uncertainty model can be used to define CPPS behavior in this case. Meanwhile, various stakeholders are involved in the CPPS operation, which necessitates them to make important decisions in the midst of uncertainty. When the different decision-making stakeholders involved in CPPSs are interacting, they each pursue their own strategy for profit based on the information available. This type of probability problem is described using game-theoretic models.

3.2.2.4.1 Uncertainty model

The uncertainty model can be utilized to characterize the behavior of CPPSs where the occurrences in physical and cyber systems are difficult to describe. Meanwhile, various stakeholders are involved in the CPPS operation, which necessitates them to make critical decisions in the face of uncertainty. If there are interactions among the various decision-making backers involved in the operation of CPPS, each of them uses the information available to pursue their own strategy for profit. This type of probability problem is described using game-theoretic models.

3.2.2.4.2 Game-theoretic model

Cyber-attacks on the physical power supply have attracted a lot of attention in the recent years. CPPS's information and communication systems were targeted by cyber-attacks, but the power system was protected by detection and mitigation of cyber-attacks. Cyber-physical interaction is described by a game-theoretic model for attacking and mitigating threats, which is used for risk analysis, vulnerability analysis, and threat analysis as well. Participating in CPPS allows defenders to make decisions in a competitive environment for their own benefit. Modeling the competitive connection among CPPS participants could be done by using a Colonel Blotto game [29], a zero-sum game [30], or a stochastic game [31].

3.2.2.5 Simulation model

The physical system's continuous nature and the cyber system's discrete components make the CPPS study problematic. The simulation model aids power system operators in achieving unified modeling of the continuous system's dynamic behavior and the discrete system's static behavior.

3.2.3 CPPS interdependent modeling

The CPPS is a large-scale, interconnected multipart system with many physical and cyber components. The dynamic interface interactions between cyber

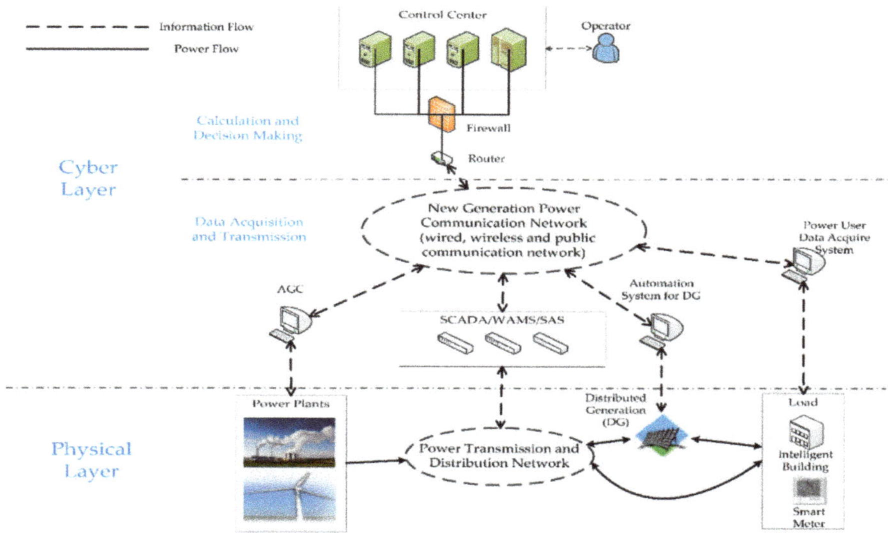

Figure 3.8 CPPS interdependent modeling.

and physical equipment are illustrated using CPPS interdependent modeling. The interdependent CPPS is separated into three distinct layers, as shown in Figure 3.8. The physical layer node represents the generation, transformers, station, and so on, while the physical layer edge represents the powerlines in the electrical power grid. The primary responsibility of the cyber layer nodes is to monitor and control the CPPS. The communication links between the cyber nodes are indicated by the network edges. One-to-one and one-to-multiple interdependencies are the two forms of interdependencies in CPPS based on cyber layer nodes [32]. The single cyber node monitors and controls each physical node in the one-to-one interdependency. The data from the dispersed cyber nodes is then collected by the command center. Each physical node is observed by numerous cyber nodes in one-to-many interdependencies, which is especially useful for safeguarding data against cyber threats.

The risk associated with smart grids increases when the robust codependent CPPS's coexisting cyber-side failure causes the physical outage's corrective control to fail. In this situation, it is necessary to deliver crucial information about the CPPS in a safe way to ensure its accessibility and increase the resilience of smart grids in the event of potential cyber–physical connection failures. In a CPS-resilient routing model, a cyber–physical sensitivity-based data flow is recommended.

3.3 Cyber-attacks and Cybersecurity in CPPS

The electrical grid is one of the most valuable assets of the country and an excellent example of a cyber–physical system. Advanced communications and information technologies are closely monitored and managed because they require close coordination of computing, telecommunication, controlling, and interaction. Despite the fact that digital technologies are more capable and dependable at monitoring and controlling it, the electrical grid is vulnerable to cyber-attacks and has a complicated interaction between cyber and physical systems. A cyber threat on the electrical power system jeopardizes the system's secure functioning by disrupting the information flow. Large-scale CPPS becomes more difficult to detect once physical outbreaks on the power system are successfully linked with cyber-attacks. Physical attacks, which tend to activate a torrential catastrophe across the electric power grid system, hide future cyber-attacks. As a result, CPPS's frequent cyber-attacks and cybersecurity measures must be examined. Cyber threats pose a substantial threat to key infrastructure, such as the electrical grid, and cybersecurity research is a top priority for most R&D organizations throughout the world.

3.3.1 Cybersecurity for CPPS

The CPPS covers data security, telecommunication infrastructure security, and vulnerability scanning. According to a previous study, traditional data future technologies (IT) security characteristics are inadequate for CPPS, and the following cybersecurity research areas for CPPS have been recognized: (i) attack-resistant tracking, safety, and controller design, (ii) protection against harmonized cyber threats, (iii) infrastructure security, and (iv) simulation models. The majority of the R&D needs for the emerging CPPS are for cyber-attack detection, prevention, mitigation, and resilience, which are all aspects of power grid cybersecurity. The main theme of smart grid cybersecurity research is to progress a unified performance modeling framework that takes physical and cyber system dynamics into account. After that, the model is used to calculate the influence of a cyber threat on the power system in terms of load loss, stability difficulties, financial losses, and equipment damage. Following the risk assessment, the power system must be protected against a variety of cyber-attacks, including intrusions, denial-of-service attacks, coordinated attacks, and malware-based attacks.

Possible attacks, their motive, and available resources could all be considered a danger. The CPPS susceptibility is evaluated by the grid's cyber sophisticated supporting infrastructure. Protocols, software, networks, and other resources are frequently used to support monitoring, controlling, and protecting activities. The CPPS has an impact when several cyber susceptibilities affect the grid's multiple power applications for controlling the physical system.

Generation, transmission, and distribution are the three key domains that make up the electricity system. Each domain is in charge of its own set of machines, protocols, and communication signals. As a result, each control system has its own set of risks, weaknesses, and implications for CPPS operations.

Attack detection, attack prevention, attack mitigation, attack forensics, and attack resilience are all part of the cybersecurity life-cycle model for attack-resistant wide-area monitoring, protection, and control systems in the electrical grid [33].

Attack deterrence: It is the ability of a protector to persuade a potential enemy to stop from assaulting.

Attack prevention: It is the ability of the defense to avoid system attacks over risk valuation, risk reduction, and cybersecurity technology, among other things.

Attack detection: In both online and offline modes, the defense must detect the attack.

Attack mitigation: The defense should employ the right mitigation method in order to keep the system operational without compromising the grid's performance, stability, or security [34].

Attack resilience: If the system is attacked, it must have sufficient resiliency to retain the system's working status, although at a reduced degree of performance, security, or stability [35].

Attack forensics: Forensic examination can help identify the perpetrator and source of a crime, as well as anticipate upcoming attacks. To conclude, each strut of the hub-and-spoke architecture focuses on unique cybersecurity measures that combine efficient technology with supporting technical tools to limit the spread of cyber-attacks across the cybersecurity lifecycle.

3.4 Communication Protocols

A user can send a block of data from one place to another by following a series of instructions called a communication protocol. Unambiguous formats are used by network nodes to deliver data effectively. Network communication between interacting network nodes is more secure when communication protocols are used.

- Internet Protocol Security (IPSec) and virtual private networks (VPNs)
- Secure Socket Layer (SSL) and Transport Layer Security (TLS)
- Application Transparent Transport Layer Security (AT-TLS)
- Kerberos
- Open Shortest Path First (OSPF) authentication
- Standard Network Management Protocol (SNMP) v3

3.4.1 IPSec and VPNs

The IPSec Working Group of the Internet Engineering Task Force (IETF) is in charge of establishing IPSec. It permits authentication, integrity, and data protection between any two IP entities. Through Internet Key Exchange, an IETF-defined key management system, and cryptographic keys, security associations can be manually or dynamically managed. With IPSec, virtual private networks (VPNs) can be built [36]. Using a secure tunnel known as a security association, a company can expand its private network across a public network, such as the Internet, using a VPN. IPSec VPNs safeguard sensitive data within a company's internal network in addition to guaranteeing secure data transfer over the Internet for intra- and inter-company communications.

3.4.2 SSL and TLS

The SSL methods offer message integrity, data encryption, and data origin authentication. X.509 certificates are also used to identify servers and clients. SSL starts with a handshake in which the server employs X.509 certificates to confirm the client's identity. Both the server and client can use the cryptographic methods used during SSL sessions. The client and server share a list of algorithms during the SSL handshake. Based on the best match between the client and server databases, these algorithms are selected.

3.4.3 Application Transparent Transport Layer Security

System SSL can be invoked in the TCP transport layer of the Communications Server. In the TCP/IP profile, the TTLS or NOTTLS parameter specifies whether Application Transparent Transport Layer Security (AT-TLS) should be supported [37]. When AT-TLS is enabled in Policy Agent, AT-TLS statements define the security features for connections that match AT-TLS rules. By deploying this policy-driven functionality underneath many existing connections, it removes the need for the application to know about encryption and decryption being achieved on its behalf. Applications that require TLS negotiation or client authentication will also be supported.

3.4.4 Kerberos

Kerberos is a client/server network verification system that leverages secret-key cryptography to offer strong validation. The foundation of Kerberos is the idea that workstations and services will interact over an insecure network. It enables one-way or two-way authentication between clients and servers. It encrypts data and stops passwords from being retyped in order to access networked services. Additionally, it stops passwords from being transmitted in clear text over networks. One can more effectively handle multiple passwords with the aid of this program.

3.4.5 OSPF Authentication

The Communications Server OSPF dynamic routing protocol uses the OSPF MD5 Authentication security protocol, as well-defined in RFC 2328, to support message confirmation and message truthfulness of OSPF routing packets [38]. OSPF MD5 Authentication ensures the accuracy of the OSPF routing network's routing tables by preventing unlawful IP resources from sending OSPF routing messages into the network without being detected. The MD5 technique is used by OMPROUTE to build a secure media access control for the routing message. This media access control is included in the routing message so that it may be authenticated by the receiver.

3.4.6 SNMPv3

It is categorized as User-based Protection Model (USM) because it assigns different levels of security to users that access managed data. The Simple Network Management Protocol (SNMPv3) architecture provides a number of security competencies, including USM for authentication and privacy, and view-based access control model for restraining access to different management information base objects on per-user basis [39]. SNMP, on the other hand, is about more than just security. It specifies an architecture for SNMP management frameworks with the goal of allowing parts of the architecture to develop over time without having to rewrite the entire structure. As a result, three major subsystems have been defined: message processing, security, and access control.

Even though there is a new message format, SNMPv3 messages created using the SNMPv1 and SNMPv2 formats, for instance, can still be supported [40]. The user-based security model can coexist with the community-based security models that were previously in use.

3.5 Case Studies on Fault Detection and Diagnosis of Intelligent Buildings

3.5.1 Building fault detection data to aid diagnostic algorithm creation and performance testing

Modifications to current commercial building control mechanisms, as well as accompanying efficiency gains, are expected to save 4%–5% of national energy use. Fault detection and diagnosis (FDD) approaches, on the other hand, are designed to detect the prevalence of functional flaws as well as their causes. Assessment methods, black boxes, and guideline approaches are among the strategies used by FDD. The absence of standard datasets and test methodologies to accurately assess their performance has long been a problem. This chapter presents the first publicly available dataset of its kind, which contains ground-truth data on the existence and absence of construction flaws [41].

3.5.1.1 Background

Buildings consume 40% of the world's energy and produce 33% of active and passive carbon emissions due to fuel combustion. In corporate buildings in the United States, more efficient operations and improved controls can save an average of 29% of energy. Building operational data is used by algorithms for automated FDD to verify the prevalence of issues and their primary causes. FDD is becoming more relevant to the construction sector as structures become more data-rich and as big data becomes more widely used. FDD is being used by building proprietors and operators who are on the cutting edge of technology adoption to save an average of 7% on their entire portfolio. A utility meter reading, HVAC system control data, smart lighting data, or monoelectric and gas information, for example, is typically obtained on a project-by-project basis, and is often subject to NDAs or other information sharing agreements. There are relatively few datasets that provide verifiable ground truth evidence on the existence and nonappearance of faults in FDD applications. Most problems go unnoticed because FDD has not yet been implemented in most structures. Historians do not disclose whether reported flaws are validated when FDD techniques are used, and fake positives and negatives can cause problems when interpreting historical documents.

3.5.1.2 Facilities and simulation tools

Using HVACSIM+ and a co-simulation between Energy Plus and Modelica, the simulated datasets were produced. The US National Institute of Standards and Technology developed HVACSIM+, the Lawrence Berkeley National Laboratory developed the Modelica Buildings Library, and Energy Plus was developed by numerous contributors with funding from the US Department of Energy. HVACSIM+, Modelica, and Energy Plus are non-hereditary modeling tools for physics-based modeling of building HVAC systems, as discussed in relation to other modeling tools in HVACSIM+. Three experimental research centers were used to construct the experimental datasets. FLEXLAB, based at the Lawrence Berkeley National Laboratory, is meant to test the effectiveness of critical building systems in real-world circumstances, either individually or as a group. Each of the facility's buildings houses a pair of identical test cells. A control and data acquisition platform from National Instruments is used to operate the plant. The dataset in this study was created using a 20-by-25-foot zone supplied by a 10-ton forced air chiller coupled with another cell.

In accordance with ASHRAE Guideline 3618, two experimental datasets were created in FLEXLAB: (i) constant air volume mode and (ii) univariate air volume mode. Both were created using National Instruments (NI) Test Stand programming environment. The Energy Resource Station facility was established by the Iowa Energy Center in Ames City, Iowa, to compare and analyze diverse energy performance indicators. The test system was controlled by a commercially existing building automation system. An AHU that served three outer zones and one inside zone was subjected to fault testing. The system includes variable air volume space cooling.

3.5.1.3 Technical validation

Sensor data, as well as data on equipment state and operational directives, are included in the facility measurements dataset. Temperature, humidity, energy, force, and air movement are all measured with sensors. Valve and damper control orders, operational modes, and temperature setpoints are among the equipment status and control directives. Sensors are calibrated at the facility on a regular basis using a number of typical methods based on ISO/IEC 1702523, which specifies standards for inspection and calibration laboratories. A complete example of FLEXLAB AHU temperature sensor water-bath calibration is provided (thermistors). The simulated functioning data under a specific fault were evaluated to make sure they accurately recreated the key fault signature collected from the facility in order to verify the model for faulty situations.

3.5.2 Edge computing and adaptive fault-tolerant tracking control algorithm for smart buildings: A case study

Internet of Things (IoT) and edge computing devices are contributing to the building of a more digital, intelligent, and connected society through their development and integration. As a result, massive amounts of data are produced as a result of a variety of human activities. In intelligent buildings, for example, tens of thousands of systems and suppliers interact. In this setting, failures of measurement and control systems are common. This work uses a class of Internet of Things-based temperature systems that are vulnerable to precision state uncertainties and external disruptions as a case study to examine fault-tolerant robust adaptive monitoring control with state prediction performance. To decrease tracking error and increase algorithm performance, the authors suggest a novel control technique based on consensus game theory and anticipating future states. The author describes how they created a new packet network measurement and management algorithm. Compared to previous results, this is aimed at increasing energy efficiency and ensuring the efficacy of adaptive temperature control. The new fault tolerance error tracking methodology assumes vigorous monitoring of the reference model according to the simulation described in this task. The expected temperature signal has been shown to be constrained by a small area near the collected temperature data.

3.5.2.1 Proposed control model

An adaptive control algorithm is presented in this work. This algorithm's main objective is to control the temperature of the smart building. To forecast the accuracy condition of the sensors over time, this approach uses an adaptive control mechanism. In order to take full advantage of temperature control, the proposed method uses three mechanisms: (1) Module for data quality: This module enriches the quality of collected IoT data by finding faulty data. (2) State forecasting: This method determines what the state error will be in the future based on the current error produced by the IoT nodes.

(3) Controller: This mechanism uses PID control algorithm, which gets the optimized inputs from previous two mechanisms.

3.5.2.2 Reference input

The reference input in this system is connected to a central unit where the building's temperature is controlled by the property manager. The actuators (thermostats) are then controlled by the algorithm to ensure that the building achieves and maintains the appropriate temperature.

3.5.2.3 State predictor and cooperative control algorithm

This stage of the algorithm predicts whether the IoT nodes are in a defective state at that instance and then outputs the anticipated temperature. In order to detect inaccurate data and increase the quality of the data obtained by the sensors, this technique compares the sensor's neighborhood temperatures using a cooperative game-based theory.

3.5.2.4 Results

Smart building data is used to test for the proposed paradigm [42]. The sensors were mounted in each construction component, and the IoT nodes were also positioned. This adaptive approach predicts the output temperature and contrasts it with the temperature obtained from the IoT network using the precise state estimate data of the adaptive algorithm. Even when disturbances are introduced, the closed-loop signals are confined by the predefined temperature interval, and the settling period is relatively short.

3.6 Conclusion

In this chapter, various modeling techniques of cyber–physical systems are discussed. Communication protocols employed in the cyber–physical system are discussed. Finally, this chapter is concluded with a case study on using cyber–physical systems for fault diagnosis techniques employed in intelligent buildings.

References

[1] E. F. Orumwense and K. Abo-Al-Ez, "A systematic review to aligning research paths: Energy cyber-physical systems," Cogent Eng., vol. 6, no. 1, pp. 1–21, Dec. 2019.

[2] L. Shi, Q. Dai, and Y. Ni, "Cyber–physical interactions in power systems: A re- view of models, methods, and applications," Electr. Power Syst. Res., vol. 163, pp. 396–412, Oct. 2018.

[3] S. Suryanarayanan, R. Roche, and T. M. Hansen, "Cyber-physical-social systems and constructs in electric power engineering," Inst. Eng. Technol., London, U.K., Tech. Rep., 2016.
[4] Y. Cao, Y. Li, X. Liu, and C. Rehtanz, Cyber-Physical Energy and Power Systems. Singapore: Springer, 2020.
[5] C. Dong, H. Jia, T. Jiang, L. Bai, Q. Hu, L. Wang, and Y. Jiang, "Effective method to determine time-delay stability margin and its application to power systems," IET Gener., Transmiss. Distrib., vol. 11, no. 7, pp. 1661–1670, May 2017.
[6] H. Bevrani, M. Watanabe, and Y. Mitani, Power System Monitoring and Control. Hoboken, NJ, USA: Wiley, 2014.
[7] C. Li, G. Li, C. Wang, and Z. Du, "Eigenvalue sensitivity and eigenvalue tracing of power systems with inclusion of time delays," IEEE Trans. Power Syst., vol. 33, no. 4, pp. 3711–3719, Jul. 2018
[8] Q. Mou, H. Ye, and Y. Liu, "Enabling highly efficient eigen-analysis of large delayed cyber-physical power systems by partial spectral discretization," IEEE Trans. Power Syst., vol. 35, no. 2, pp. 1499–1508, Mar. 2020.
[9] H. Ye, Q. Mou, and Y. Liu, "Calculation of critical oscillation modes for large delayed cyber-physical power system using pseudo-spectral discretization of solution operator," IEEE Trans. Power Syst., vol. 32, no. 6, pp. 4464–4476, Nov. 2017.
[10] H. Ye, T. Li, and Y. Liu, "Time integration-based IGD methods for eigen analysis of large delayed cyber-physical power system," IEEE Trans. Power Syst., vol. 35, no. 2, pp. 1376–1388, Mar. 2020.
[11] W. Gao, H. Ye, Y. Liu, L. Wang, and W. Ci, "Comparison of three stability analysis methods for delayed cyber-physical power system," in Proc. China Int. Conf. Electr. Distrib. (CICED), Xi'an, China, pp. 1–5, 2016.
[12] S. Xin, Q. Guo, H. Sun, B. Zhang, J. Wang, and C. Chen, "Cyber-physical modeling and cyber-contingency assessment of hierarchical control systems," IEEE Trans. Smart Grid, vol. 6, no. 5, pp. 2375–2385, Sep. 2015.
[13] Z. Su, S. Xin, L. Xu, W. Li, Z. Shi, Q. Guo, and H. Sun, "A compensation method based assessment of cyber contingency for cyber-physical power systems," in Proc. IEEE Power Energy Soc. Gen. Meeting (PESGM), Portland, OR, USA, pp. 1–5, 2018
[14] K. Chen, F. Wen, and I. Palu, "Cyber contingencies impacts analysis in cyber physical power system," in Proc. IEEE Int. Conf. Energy Internet (ICEI), Nanjing, China, pp. 37–41, 2019.

[15] G. Cao, W. Gu, P. Li, W. Sheng, K. Liu, L. Sun, Z. Cao, and J. Pan, "Operational risk evaluation of active distribution networks considering cyber contingencies," IEEE Trans. Ind. Informat., vol. 16, no. 6, pp. 3849–3861, Jun. 2020.
[16] S. Zonouz, C. M. Davis, K. R. Davis, R. Berthier, R. B. Bobba, and W.H. Sanders, "SOCCA: A security-oriented cyber-physical contingency analysis in power infrastructures," IEEE Trans. Smart Grid, vol. 5, no. 1, pp. 3–13, Jan. 2014.
[17] C.-W. Ten, A. Ginter, and R. Bulbul, "Cyber-based contingency analysis," IEEE Trans. Power Syst., vol. 31, no. 4, pp. 3040–3050, Jul. 2016.
[18] G. Wu, G. Wang, J. Sun, and J. Chen, "Optimal partial feedback attacks in cyber-physical power systems," IEEE Trans. Autom. Control, early access, 2020.
[19] L. Wang, Z. Qu, Y. Li, K. Hu, J. Sun, K. Xue, and M. Cui, "Method for extracting patterns of coordinated network attacks on electric power CPS based on Temporal–Topological correlation," IEEE Access, vol. 8, pp. 57260–57272, 2020.
[20] K. R. Davis, C. M. Davis, S. A. Zonouz, R. B. Bobba, R. Berthier, L. Garcia, and P. W. Sauer, "A cyber-physical modeling and assessment framework for power grid infrastructures," IEEE Trans. Smart Grid, vol. 6, no. 5, pp. 2464–2475, Sep. 2015.
[21] P. A. Khand, "System level security modeling using attack trees," in Proc. 2nd Int. Conf. Comput., Control Commun. (IC4), pp. 1–6, 2009.
[22] P. A. S. Ralston, J. H. Graham, and J. L. Hieb, "Cyber security risk assessment for SCADA and DCS networks," ISA Trans., vol. 46, no. 4, pp. 583–594, Oct. 2007.
[23] C. W. Ten, C. C. Liu, and M. Govindarasu, "Vulnerability assessment of cybersecurity for SCADA systems using attack trees," in Proc. IEEE Power Eng. Soc. Gen. Meeting PES, pp. 1–8,2007.
[24] Y. Wadhawan, A. AlMajali, and C. Neuman, "A comprehensive analysis of smart grid systems against cyber-physical attacks," Electronics, vol. 7, no. 10, p. 249, Oct. 2018.
[25] N. Liu, J. Zhang, H. Zhang, and W. Liu, "Security assessment for communication networks of power control systems using attack graph and MCDM," IEEE Trans. Power Del., vol. 25, no. 3, pp. 1492–1500, Jul. 2010.
[26] Y. F. Wang, K. L. Gao, T. Zhao, and J. Qiu, "Assessing the harmfulness of cascading failures across space in electric cyber-physical system based on improved attack graph," Proc. CSEE, vol. 36, no. 6, pp. 1490–1499, 2016.

[27] S. Zonouz, C. M. Davis, K. R. Davis, R. Berthier, R. B. Bobba, and W. H. Sanders, "SOCCA: A security-oriented cyber-physical contingency analysis in power infrastructures," IEEE Trans. Smart Grid, vol. 5, no. 1, pp. 3–13, Jan. 2014.
[28] Y. Li, Y. Wang, and S. Hu, "Online generative adversary network based measurement recovery in false data injection attacks: A cyber-physical approach," IEEE Trans. Ind. Informat., vol. 16, no. 3, pp. 2031–2043, Mar. 2020.
[29] Y. Zhang, L. Wang, Y. Xiang, and C.-W. Ten, "Inclusion of SCADA cyber vulnerability in power system reliability assessment considering optimal resources allocation," IEEE Trans. Power Syst., vol. 31, no. 6, pp. 4379–4394, Nov. 2016.
[30] M. Husák, J. Komárková, E. Bou-Harb, and P. Celeda, "Survey of attack projection, prediction, and forecasting in cyber security," IEEE Commun. Surveys Tuts., vol. 21, no. 1, pp. 640–660, 1st Quart., 2019.
[31] A. Anwar, A. N. Mahmood, and Z. Tari, "Identification of vulnerable node clusters against false data injection attack in an AMI based smart grid," Inf. Syst., vol. 53, pp. 201–212, Oct. 2015.
[32] R. V. Yohanandhan and L. Srinivasan, "Decentralised wide-area fractional order damping controller for a large-scale power system," IET Gener., Transmiss. Distrib., vol. 10, no. 5, pp. 1164–1178, Apr. 2016.
[33] L. Wei, A. I. Sarwat, W. Saad, and S. Biswas, "Stochastic games for power grid protection against coordinated cyber-physical attacks," IEEE Trans. Smart Grid, vol. 9, no. 2, pp. 684–694, Mar. 2018.
[34] M. Esmalifalak, G. Shi, Z. Han, and L. Song, "Bad data injection attack and defense in electricity market using game theory study," IEEE Trans. Smart Grid, vol. 4, no. 1, pp. 160–169, Mar. 2013.
[35] A. Ashok, M. Govindarasu, and J. Wang, "Cyber-physical attack-resilient wide-area monitoring, protection, and control for the power grid," Proc. IEEE, vol. 105, no. 7, pp. 1389–1407, Jul. 2017.
[36] A. Patel and S. Purwar, "Switching attacks on smart grid using nonlinear sliding surface," IET Cyber-Phys. Syst. Theory Appl., vol. 4, no. 4, pp. 382–392, Dec. 2019.
[37] H. M. Khalid, S. M. Muyeen, and J. C.-H. Peng, "Cyber-attacks in a looped energy-water nexus: An inoculated sub-observer-based approach," IEEE Syst. J., vol. 14, no. 2, pp. 2054–2065, Jun. 2020.
[38] Q. Wang, W. Tai, Y. Tang, and M. Ni, "Review of the false data injection attack against the cyber-physical power system," IET Cyber-Phys. Syst. Theory Appl., vol. 4, no. 2, pp. 101–107, Jun. 2019.

[39] L. Yu, X.-M. Sun, and T. Sui, "False-data injection attack in electricity generation system subject to actuator saturation: Analysis and design," IEEE Trans. Syst., Man, Cybern. Syst., vol. 49, no. 8, pp. 1712–1719, Aug. 2019.

[40] S. Ahmadian, X. Tang, H. A. Malki, and Z. Han, "Modelling cyber attacks on electricity market using mathematical programming with equilibrium constraints," IEEE Access, vol. 7, pp. 27376–27388, 2019, doi: 10.1109/ACCESS.2019.2899293.

[41] J. Granderson, G. Lin, A. Harding, P. Im, and Y. Chen, "Building fault detection data to aid diagnostic algorithm creation and performance testing", Science Data, vol. 7, 2020.

[42] R. Vara, I. Candanedo, F. Prieta, S. Rodríguez, J. L. Calvo-Rolle, G. K. Venayagamoorthy, P. Vega, and J. Prieto, "Edge Computing and Adaptive Fault-Tolerant Tracking Control Algorithm for Smart Buildings: A Case Study", Cybernetics and Systems, vol. 51, 2020.

4

Diagnosis of Power Components Amid the 4th Industrial Revolution

Mahmoud Elsisi[1,2], Minh-Quang Tran[3,4], Mohammed Amer[5], Viet Q. Vu[4], Mohamed M. F. Darwish[2,6], Karar Mahmoud[6,7], Diaa-Eldin A. Mansour[8,9], and Matti Lehtonen[6]

[1]Department of Electrical Engineering, National Kaohsiung University of Science and Technology, Taiwan
[2]Department of Electrical Engineering, Faculty of Engineering at Shoubra, Benha University, Egypt
[3]Department of Mechanical Engineering, TUETECH University, Vietnam
[4]Department of Mechanical Engineering, Thai Nguyen University of Technology, Vietnam
[5]Department of Mechanical Engineering, Palestine Technical University – Kadoorie, Palestine
[6]Department of Electrical Engineering and Automation, School of Electrical Engineering, Aalto University, Finland
[7]Department of Electrical Engineering, Faculty of Engineering, Aswan University, Egypt
[8]Department of Electrical Power and Machines Engineering, Faculty of Engineering, Tanta University, Egypt
[9]Department of Electrical Power Engineering, Faculty of Engineering, Egypt-Japan University of Science and Technology (E-JUST), Egypt

Abstract

Power components are among the most expensive parts in the power and industrial sectors worldwide. Nevertheless, fault diagnosis and remote control still face a lot of challenges in providing high accuracy and fast response to protect the power components. The upgrade of the Internet of Things (IoT) in recent years has made apps smarter, and linked gadgets make them more useful in smart systems. In the meantime, IoT and machine learning (ML) are being used to give an application more intelligence and capabilities, especially since the amount of data collected is increasing. As the trend of IoT and machine learning has just begun worldwide, especially in electrical applications, it seemed important to present a comprehensive overview of the current actions and trends, including the regulatory limits and reservations about smart energy systems. In this sense, this chapter is concerned with smart energy systems, which have been investigated using both machine learning and the Internet of Things. The aim of this chapter is to demonstrate recent advances in IoT and machine learning to troubleshoot smart energy systems. The chapter deals with fault diagnosis and remote control for smart energy systems based on advanced digital twin platforms and machine learning techniques that are more suitable for modern smart grids and fourth-generation (4G) and fifth-generation (5G) networks. Following the trend of the fourth industrial trend (named Industry 4.0), the use of online condition monitoring systems has accelerated the process of fault diagnosis for smart energy systems. Finally, in support of Industry 4.0, several case studies covering online fault diagnosis and cybersecurity are presented, including gas-insulated switchgear, power transformers, and induction machines.

4.1 Configuration of IoT Architecture

A 5G network and cloud computing can increase system performance by providing fast and dynamic infrastructure; for instance, in complicated and large-scale systems that rely on feedback control as well as IoT [1]. Using cloud computing in industrial control applications, it is possible because of its high speed and prominent features. Many problems associated with physical systems can be overcome with cloud computing [2]. Today, many industrial vendors offer hardware and software tools designed specifically for IoT applications. These services offer data analytics architectures for a variety of industrial products, as illustrated in Figure 4.1 [3]. For objectives such as analysis storage and visualization, measurements from the sensors perform the data acquisition process. Various communication interface protocols like Modbus, open platform communications, and network protocols similar to Hypertext Transfer Protocol (HTTP) and message queuing telemetry transport (MQTT) are used to acquire the data. A variety of IoT companies are currently available for collecting data in the scope edge computing systems plus cloud gateways. Selecting the right method of data acquisition impacts the performance of real-time monitoring. When analyzing data in a complicated latency manner, offline data processing is recommended. Common

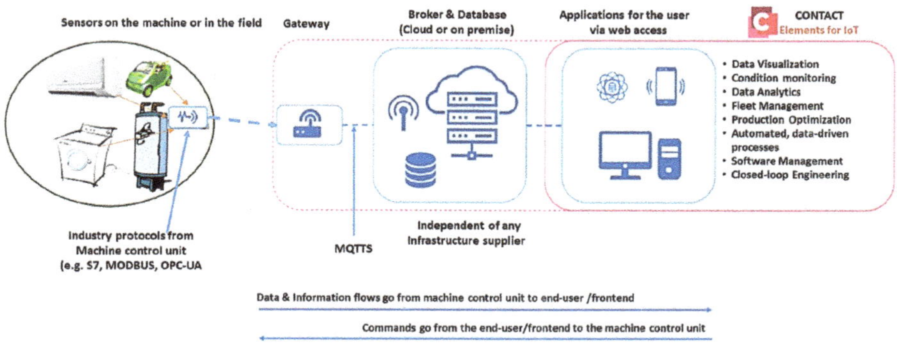

Figure 4.1 An illustration of the IoT paradigm.

data pre-processing procedures include data visualization, data selection, and data filtering. Data pre-processing can also include delay estimation, dimension reduction, and resampling. During the process of data analysis, the choice of the proper suitable model, along with training, testing, and validation data, is crucial. Note that each grid has its own environmental characteristics that determine which data analysis model should be used. Depending on edge computing as a resilient and synchronized computing construction, the proposed system is constructed. By using several communication protocols for the sensors, the devices collect data from the sensors on the machines and send it to a database server. Data collected by the gateway is classified using machine learning techniques, and the results are sent through the MQTT protocol to be viewed on the dashboard of the industrial IoT platform.

Smart sensors and microcontrollers used in Industry 4.0 are equipped with real-time data exchange capabilities to provide real-time data from all kinds of sources, including microgrids, factories, homes, and hospitals. In fact, IoT-enabled smart microcontrollers benefit from connected networks, like Bluetooth and global system for mobile communication, that provide access to examining and managing them remotely. By using technologies like IoT and big data to interface with increasingly smart devices and users, the electric vehicle (EV) industry has taken advantage of the development of Industry 4.0, 4G, and 5G networks. Therefore, a number of frameworks and platforms are offered under the umbrella of the Industry 4.0 initiative to specific industries using various technologies. Connected devices can pose security risks as a result of the variety of IoT technologies. Consequently, it represents a major challenge that faces the growth of IoT applications in global power systems technologies.

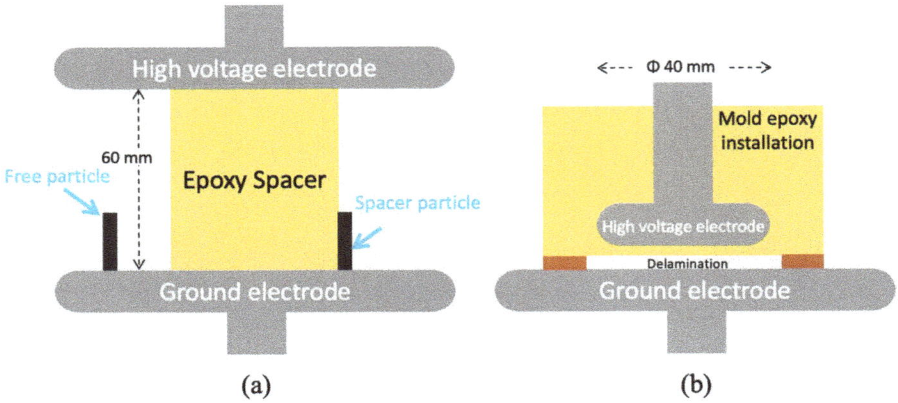

Figure 4.2 The most familiar PD faults in GIS: (a) free and spacer particle faults (MPG and MPS); (b) delamination defects (EID).

This chapter demonstrates the use of IoT frameworks under Industry 4.0, 4G, and 5G networks to monitor power components to facilitate the manufacturing of electric products, which can improve remote monitoring, real-time visibility, forecasting, cyber–physical security, and alerts. Hence, IoT technology is expected to simplify operations and supervisory control and speed up problem-solving in large-scale manufacturing operations. Accordingly, several case studies covering online fault diagnosis and cybersecurity are presented, including gas-insulated switchgear, power transformers, and induction machines.

4.2 Gas-insulated Switchgear Features Extraction

Three different defects of gas-insulated switchgear (GIS) were constructed experimentally, as displayed in Figure 4.2. GIS defects that are pictured in Figure 4.2 are among the most common [4, 5]. Among them are metal particles within the gas gap (named free particles or MPGs), metal particles adhering to the spacer surface (named spacer particles or MPSs), as well as delamination at the electrode/insulator interface (named EIDs). Figure 4.3(a) shows free particles and spacers with a total size of 5 mm and an average diameter of 0.25 mm, while Figure 4.3(b) shows a delamination defect with a diameter of 40 mm and a depth of 50 m. The free particle and spacer particles in Figure 4.3(a) have a length value equal to 5 mm plus a diameter of 0.25 mm, while the delamination fault in Figure 4.3(b) has a diameter of 40 mm and a depth of 50 m. Each defect was sequentially performed in a pressurized GIS chamber where partial discharge (PD)

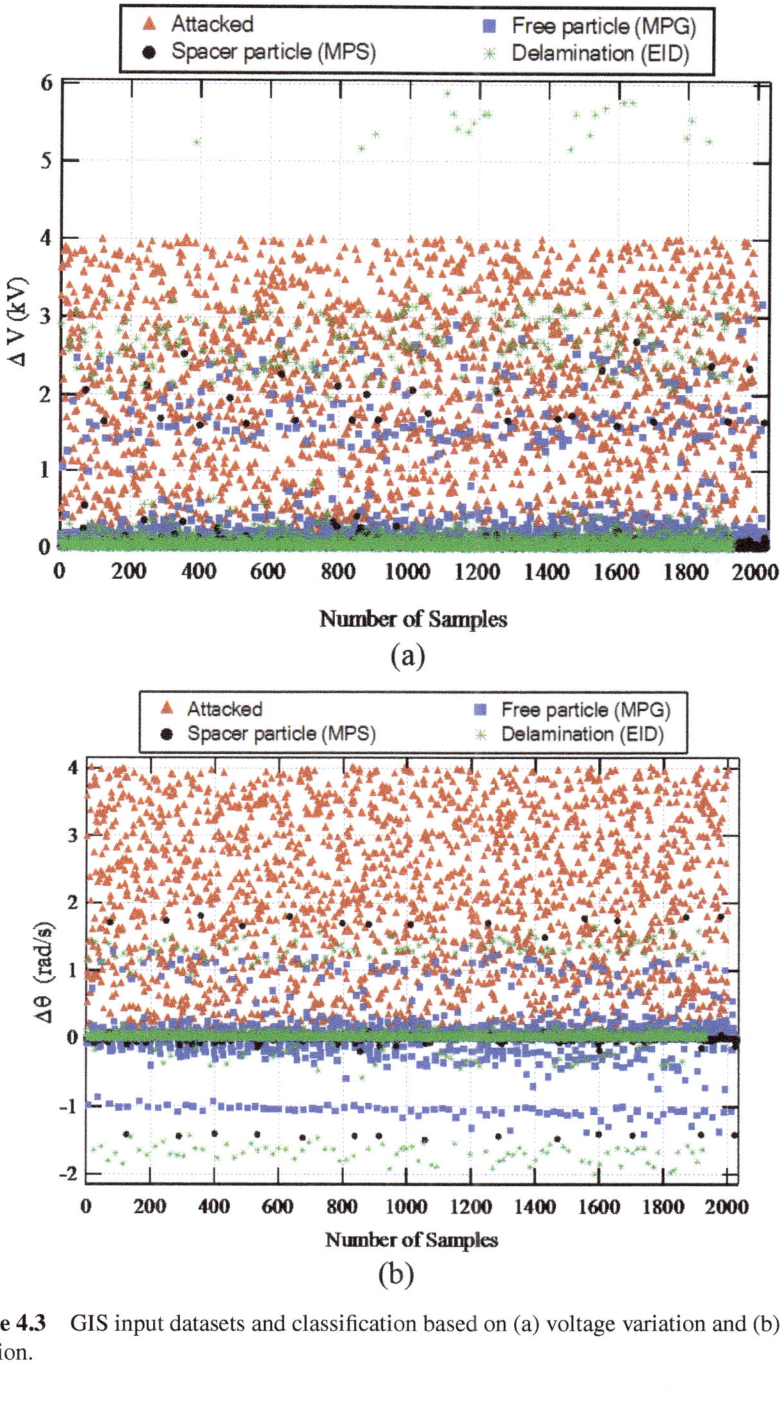

Figure 4.3 GIS input datasets and classification based on (a) voltage variation and (b) angle variation.

measurements were accomplished via current pulse measurements. After acquiring PD pulses for various faults for a time equal to 10 minutes, numerous PD features were obtained [6, 7]. Phase appearance, amplitude, pulse number, and instantaneous voltage are among the most prominent characteristics of PD. As far as PD amplitude is concerned, it is usually dependent on the defect size and needs proper calibration. It is therefore possible to misdiagnose PD based on PD amplitude. PD pulses must be calculated statistically based on PD events. As an alternative, proposed pulse sequence features, such as phase appearance as well as voltage magnitude, can be used in PD diagnosis, as previous research has proven their effectiveness [8, 9]. Further, pulse train features can be simply analyzed with voltage sensors, allowing them well suited for IoT architectures. This chapter examines four types of PD in gas-insulated switchgear utilizing the eXtreme gradient boosting (XGBoost) classifier, including the MPG, MPS, EID defect types, and cyber-attacks. For the training and testing of the XGBoost algorithm, a real-time dataset is collected from the GIS at several operating conditions. Together with the real-time GIS data, a cyber-attack dataset is generated. XGBoost classifier is trained and tested on real MPG, MPS, and EID defects, with 0 representing "attacked data" and 1, 2, and 3 representing "real data." A sample of the input dataset is shown in Figure 4.3. The dataset contains 7986 samples with four characteristics, including training dataset that accounts for 80% of the samples, and the remaining 20% of samples represent the testing dataset.

4.3 Power Transformer Features Extraction

Conventional methods for analyzing dissolved gases include taking oil samples from a transformer and measuring the concentration of different gases in specialized laboratories using gas chromatography. As a result, none of these methods provide rapid action, with oil sampling usually carried out every 3 or 6 months on a periodic basis. Furthermore, this conventional method is insufficient to observe the dynamics and trends of dissolved gases. In response, online gas sensing has gained considerable attention since it can provide continuous monitoring while being reliable.

Various online gas sensing techniques have been proposed. It has been demonstrated in [10] that near-infrared to infrared spectroscopy can be used for online gas analysis, in which six gases can be identified with a spectrum-wide dimension of 7000–667 cm^{-1}. Despite this, this approach was unable to distinguish the hydrogen (H_2) gas, which is crucial in sensing partial discharges. Online gas sensing has also been accomplished using photoacoustic spectroscopy [11–15]. Using this method, a gas's fingerprint absorption band is used to identify the type of gas, and its concentration is determined by the amount of light absorbed. Despite photoacoustic spectroscopy's advantages, it is highly susceptible to noise from surroundings. In their place, gas sensors based on semiconductor oxides and nanotechnology have been created to detect several hydrocarbon gases [16, 17]. One of the gas sensors developed for online applications is the tin dioxide stannic oxide (SnO_2) [18]. Initially, it was hypersensitive to methane (CH_4) gas; thus, it also showed sensitivity to other combustible gases. Additionally, a molybdenum disulfide (MoS_2) monolayer doped with palladium (Pd) was employed for detecting dissolved gases [19]. In tests involving H_2 and acetylene (C_2H_2) gas sensing, Pd-MoS_2 showed good performance but had weak sensitivity to CH_4. An online detector of solid oxide fuel cell (SOFC) sensor with integrated gas chromatography has also been developed [20]. Five gases could be detected – H_2, CH_4, ethane (C_2H_6), ethylene (C_2H_4), and C_2H_2 – which are widely utilized for fault diagnosis and monitoring. In the aforementioned study, gas chromatography was used to measure the dissolved gases. Therefore, the SOFC detector can be regarded as an online detector that can be well suited to inclusion in IoT infrastructures.

Electrical power transformers can be visually identified as normal operating conditions as long as the concentrations of ($H_2 < 100$ ppm, $C_2H_6 < 65$ ppm, $CH_4 < 120$ ppm, $C_2H_2 < 1$ ppm, and $C_2H_4 < 50$ ppm) are in accordance with the IEEE standard (C57.104) [21]. The if-statement condition can be used to interpret this condition. Power transformer oil contains six fault types counting PD fault, high-energy discharge (HED) fault, as well as low-energy discharge (LED) fault, besides low, medium, and high thermal (LT, MT, and HT) faults, which all need effective classifications to be detected based on their concentrations (H_2, C_2H_4, C_2H_6, CH_4, and C_2H_2) and are outlined in Table 4.1. There are 448 datasets collected and categorized into three categories of electrical faults: 48 samples are categorized as PD faults, 84 cases are categorized as LED faults, and 129 samples are categorized as HED faults. Then, a total of 91 samples were collected for the LT fault, besides 32 samples for the MT fault, and finally 64 samples for the HT fault. Therefore, the deep one-dimensional convolutional neural network (1D-CNN) proposed here is dedicated to detecting the above-mentioned six classes in the oil sample of the power transformer. A practical dataset is gathered from the power transformer oil at numerous time periods for both training as well as testing the developed 1D-CNN. Based on the input data, six classes (LT, MT, HT, LED, HED, and PD) were identified (H_2, C_2H_4, C_2H_6, CH_4, and C_2H_2) as presented in Figure 4.4.

Table 4.1 Gas values at fault cases.

Concentrations of gases with (ppm)					Fault	
H_2	CH_4	C_2H_6	C_2H_4	C_2H_2	Label	Type
105	23	13	4	3	1	PD
124	14	4	0.001	13	2	LED
235.46	333.59	177.52	1201.85	148.87	3	HED
17	21	118	0.001	4	4	LT
14	15	42	73	0.001	5	MT
100	200	110	670	11	6	HT

4.4 Fault Feature Extraction for Induction Machines

Various faults of the induction machine are investigated in this section, under different operating conditions of the motor, by means of an experimental design and data collection. Figure 4.5 depicts a motor-driven rotary system incorporating an induction motor of the type of advanced extremely high frequency, a variable frequency drive of the A510 series, and a magnetic particle brake with code ZKB010AA. To manage the torque values that are applied to the motor, a Lorenz Dr-2477-P torque meter is utilized. An accelerometer with a type of Wilcoxon 786A was used for processing the vibration signal, which was sampled at 25,600 Hz during testing. The work considers not only the normal condition, outer ring defect, as well as inner ring defect but also the cyber-attack represented by fake data as a critical condition monitoring for induction motors in real time. Figure 4.6 illustrates the outer and inner ring defects caused by machining grooves with a depth equal to 0.2 mm and a width of 0.2 mm produced by wire electrical discharge machining.

The induction motor operations were evaluated using six types of experiments with three varying bearing requirements, which included two various rotational speeds, three selections of electricity, as well as three selections of resistance. For the purpose of capturing vibration signals under numerous bearing requirements, involving the inner ring and outer ring faults, as well as a normal one, a LabVIEW program was developed. Upon reaching a steady state, the vibration signal was recorded. All bearing conditions were collected for a time span of 20 s and repeats were made three times for every condition. Eqn (4.1) and (4.2) can be utilized to describe the bearing faults' characteristic frequencies.

$$f_{\mathrm{ORF}} = \frac{N_B}{2} f_R \left(1 - \frac{D_B \cos(\theta)}{D_P}\right) \tag{4.1}$$

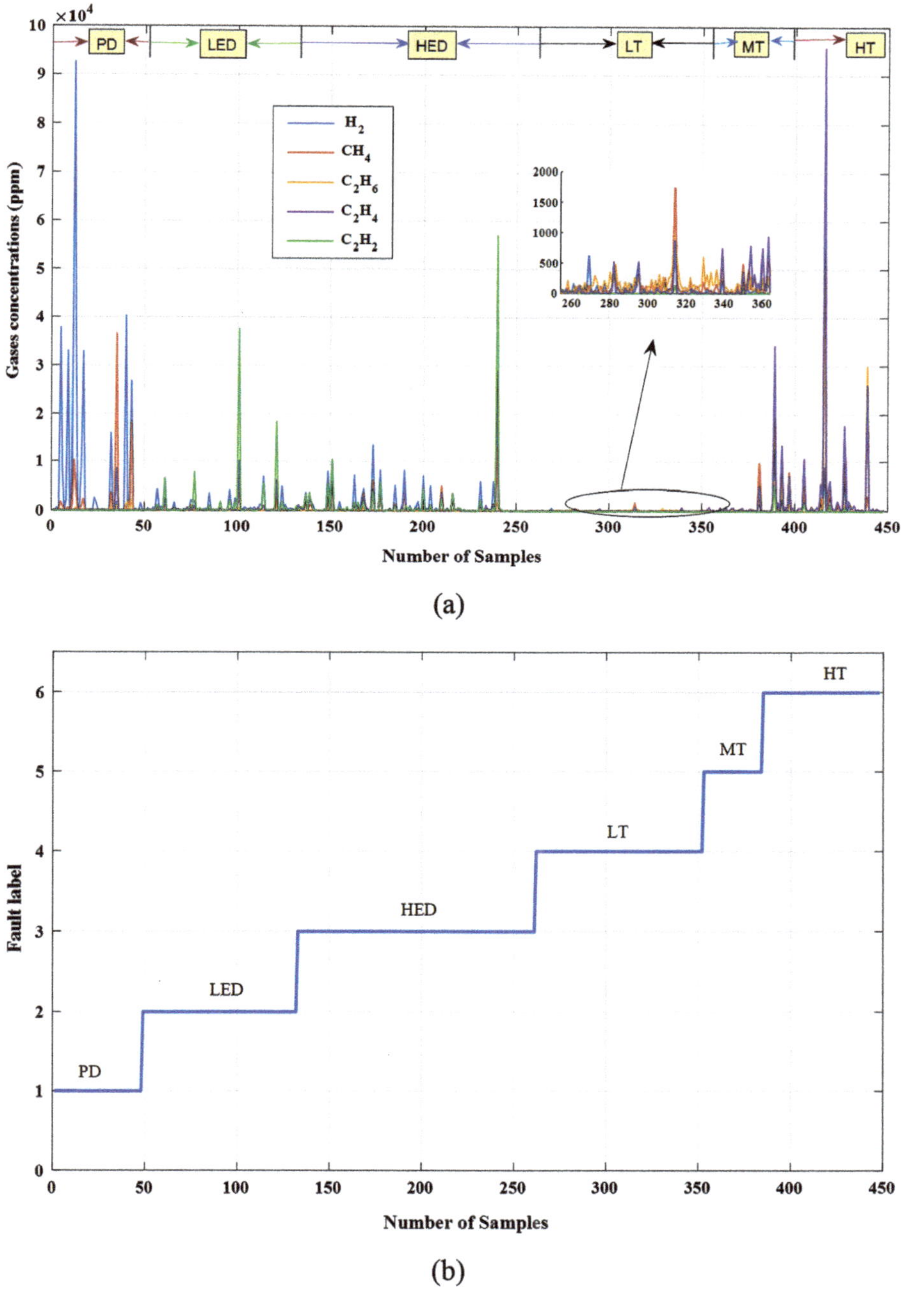

Figure 4.4 Gas concentration dataset and their fault label: (a) gas concentrations; (b) fault label.

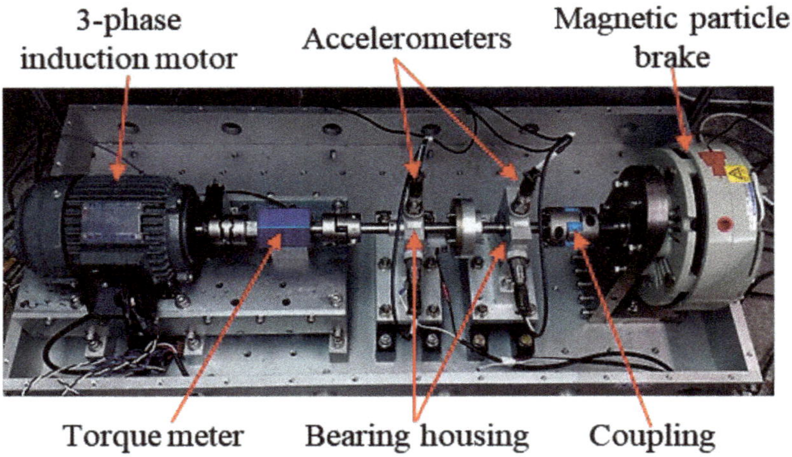

Figure 4.5 Motor-driven rotary system experimental setup.

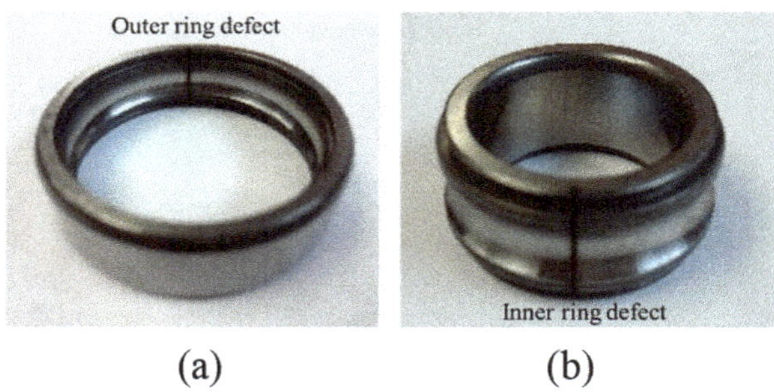

Figure 4.6 Bearing defects at (a) outer ring as well as (b) inner ring.

$$f_{\mathrm{IRF}} = \frac{N_B}{2} f_R \left(1 + \frac{D_B \cos(\theta)}{D_P}\right) \tag{4.2}$$

where f_{ORF} and f_{IRF} indicate the characteristic frequencies of the outer and inner bearing faults, respectively. N_B represents the ball number, and depicts the pitch diameter, followed by that represents the ball diameter, plus θ defined as the contact angle, and finally f_R refers to the rotation speed.

A roller bearing model 6204-T1 was employed in the experiment process, whose dimensions and parameters are described in [22]. Every experiment was analyzed by collecting and labeling the vibration signals for each bearing condition. As shown in Figure 4.7, the measured vibration signals in the time-domain spectrum show the bearing requirements under which the signals were measured. The results indicate that vibration amplitude

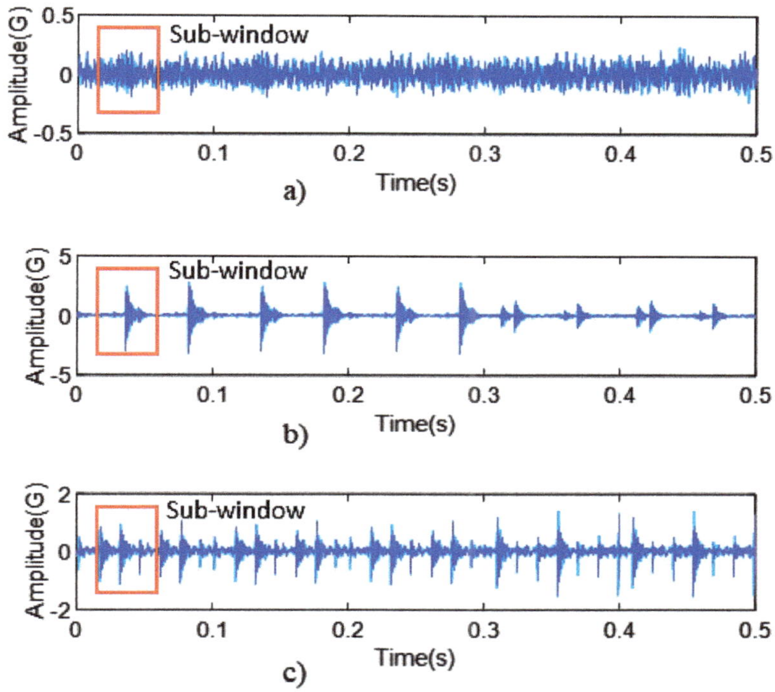

Figure 4.7 Various conditions for measuring vibration signals: (a) healthy bearing, (b) inner ring defect, as well as (c) outer ring defect.

is significantly higher during bearing failure than during healthy bearing responses. There are also different periods of shockwaves produced by a collision of the balls as well as a defect in the rings in the conditions of the inner and outer rings, as displayed in Figures 4.7(b) and (c), respectively.

4.5 Machine-learning-based Fault Diagnosis

The use of machine learning approaches for engineering applications has grown in recent years [23–26], which in turn has contributed to the diagnosis of faults with electrical components. With these approaches, the computation time can be reduced while the classification accuracy stays high. These approaches are superior to linear models [27–30]. Several methods of machine learning can be applied to classification problems. Hence, decision tree (DT), XGBoost, random forest (RF), and 1D-CNN are among the most effective ensemble methods of machine learning approaches that can be used to distinguish several fault conditions. With the appropriate data,

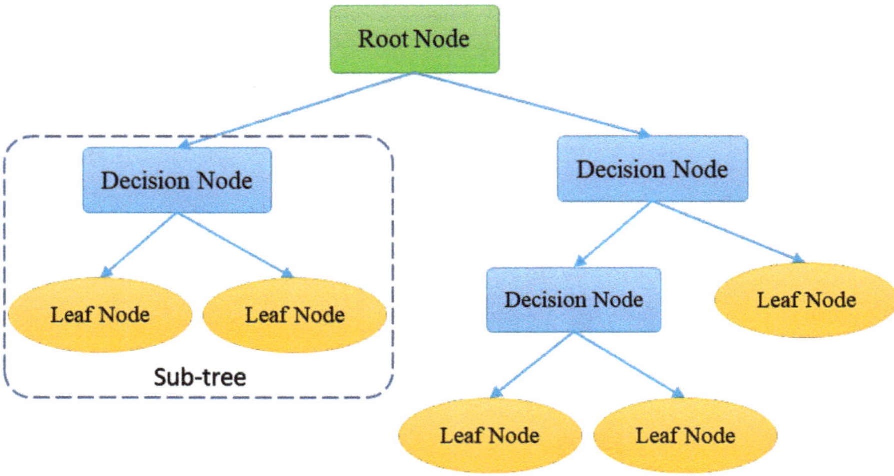

Figure 4.8 Decision tree structure.

particularly the time series, those algorithms can perform very well. The algorithms deliver high accuracy while operating at high speed. Additionally, by employing three different approaches, overfitting problems can be reduced, making them more suitable to apply to fault detection than support vector machine algorithms [31].

4.5.1 Decision tree

One of the supervised learning algorithms is the decision tree method. The method constructs a training model based on learning simple decision rules to estimate the target by examining unseen samples [32, 33]. Fast training and low memory requirements are features of the DT algorithm. When estimating the class of a dataset, the root attribute values are evaluated to the values of the real dataset attributes as part of the training process. Continually, the algorithm compares the value of the attribute with those of the sub-nodes in the following node and then moves forward. When the algorithm goes to the leaf node, the process will be ended. Figure 4.8 illustrates the schematic of a decision tree algorithm.

Suppose that the training set is symbolized with S and that the target has attributes of n different values. Then, the training set entropy S can be defined by the following equation:

$$\text{Entropy}(S) = \sum_{i=1}^{n} p_i \log_2^{p_i} \quad (4.3)$$

where p_i is the proportional value of the ith attribute specified value.

The information gain associated with an attribute that pronounces the reduction in predictable entropy is described by the following equation:

$$\text{Gain}(S, V) = \text{Entropy}(S) - \sum_{a \in A(V)} \frac{|S_a|}{|S|} \text{Entropy}(S_a) \tag{4.4}$$

where $A(V)$ represents the variety of the attribute A and S_a labels the sample set that contains the value of a for the attribute V.

Time-series datasets are used to represent the vibration dataset points over a particular time period as a feature vector or attribute. By referring to eqn (4.4), the proposed decision tree model defines an attribute within the attribute vector that has extreme information to define a bearing status based on the information gained. Nodes are selected according to the attribute with the greatest information gain. As a result, the value of entropy can be reduced from the specified root node into the leaf node.

4.5.2 Random forest

RF is considered an algorithm that is effectively used for fault diagnosis. The RF utilizes tree-type classifiers as part of an ensemble approach. By suppressing overfitting with the help of bagging, this method improves model performance [34]. The decisions of the RF are driven by the overall votes cast by the component predictors for each target. Hence, the random forest procedure construction for ensemble decision-making is illustrated in Figure 4.9.

Random forest algorithms combine multiple decision trees. Each decision tree, however, is now created using a randomly selected subset of attributes and trained against a randomly selected subset of the training data. Average voting is then made from each separate decision tree to predict the random forest model. By doing so, the RF algorithm helps to decrease overfitting among decision trees. RF begins with a choice of random cases from a training set of vibration data; after that, a restricted number of attributes are generated arbitrarily within a subset of data to make up the random attribute vector. In accordance with the samples in the collection, the decision tree is created where each single decision tree has its own individual decision. In the end, the prediction output with the maximum score is selected as the output by applying a majority voting process.

4.5.3 Extreme gradient boosting

A scalable end-to-end boosting scheme called eXtreme gradient boosting was first introduced by Chen *et al.* [35]. The fault classification problem solution

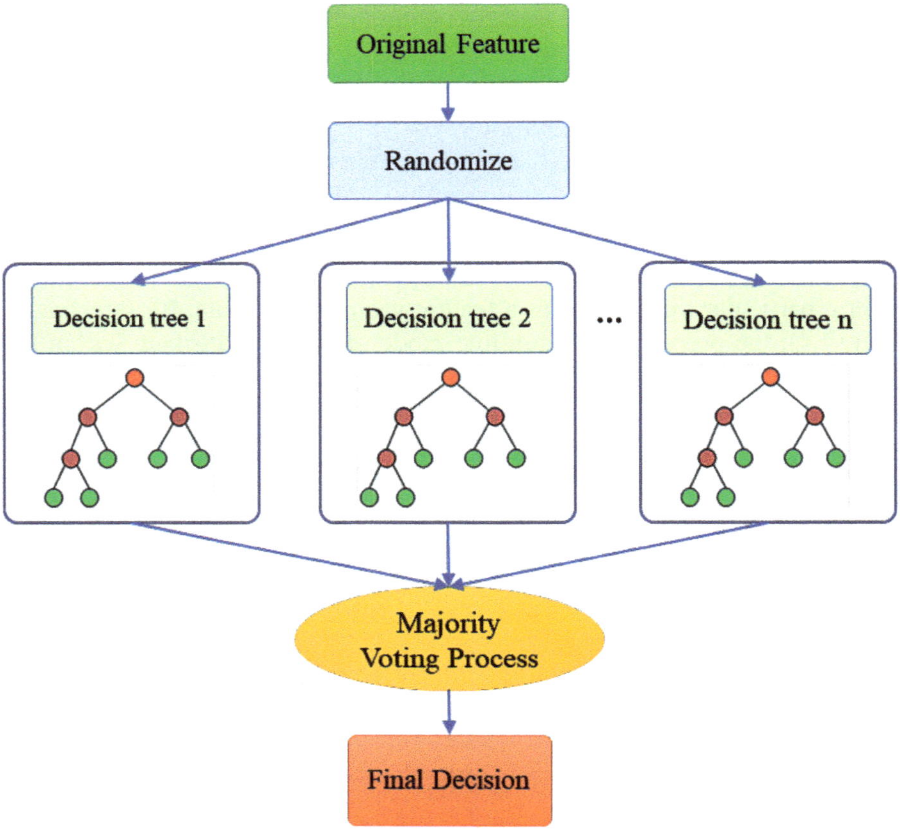

Figure 4.9 An ensemble decision-making model based on the random forest structure.

XGBoost has been widely used since it has great performance [31, 36]. By using the gradient boosting procedure, novel models can be generated to evaluate previous models' errors. Yet, a stochastic gradient descent algorithm reduces the loss in the final decision by combining the two models. Stochastic gradient descent is one of the most popular algorithms for optimizing an objective function iteratively [37]. With the help of XGBoost, computational resources for the boosted tree algorithms can be stretched to their limits, enabling a reduction in calculations and an improvement in classification speed. XGBoost is particularly effective at suppressing the overfitting problem since it simplifies the objective functions. As part of the XGBoost procedure, the first learner is close-fitting to all of the data before

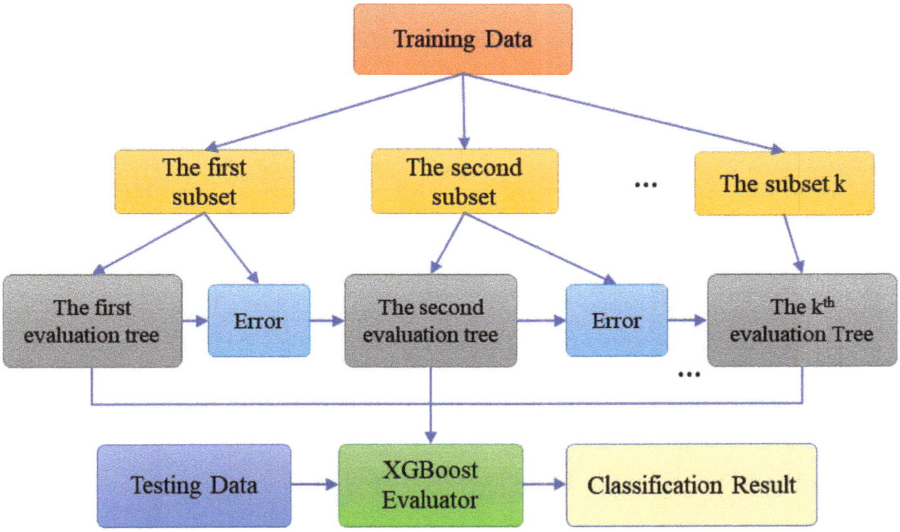

Figure 4.10 Scheme of eXtreme gradient boosting classifier.

it continues to iterate. The second learner will then fix the error of the first learner. Learning will continue until a stopping condition is met, upon which it will be considered complete. Figure 4.10 demonstrates how the XGBoost classifier works for detecting motor faults.

4.5.4 1D convolutional neural network

Among the deep convolutional neural networks, the 1D convolutional neural network stands particularly efficient for manipulating 1D data in one dimension [38]. CNN models typically consist of two types of layers: convolutional layers and fully connected layers [39]. Note that convolutional layers operate by sliding one Kernel over input data to get new features. Before passing to the fully connected layer, the convolutional layer output will be compressed. At that point, predictions of the input class will be made in the output layer. Figure 4.11 illustrates the typical architecture of a 1D convolutional neural network, which involves an input layer, followed by a convolution layer, then a pooling layer and flattening step, and finally the fully connected layer.

The data is received in one dimension with values of n cases besides c factors. Note that input data is received in one dimension with various n cases besides c factors. Here, the output layer consists of a multilayer perceptron

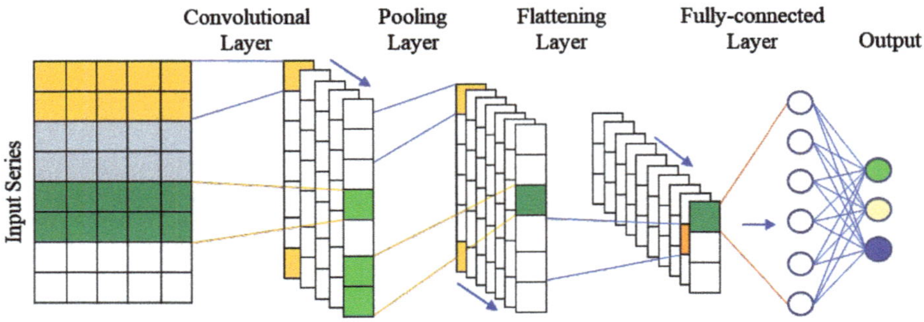

Figure 4.11 A proposed 1D convolution neural network construction.

whose size of neurons is proportional to the size of required targets [40]. Eqn (4.5) gives a description of the 1D convolution:

$$x_j^k = f\left(\sum_{i=1}^{N} x_i^{k-1} \cdot K_{\text{ij}}^k + \text{w}_j^k\right) \tag{4.5}$$

where x denotes the feature map input, w depicts the bias, K denotes the convolutional kernels, j describes the dimension of the kernels, i denotes the ith neurons at the kth layer, f means the activation formula, and, finally, (•) symbolizes the convolution factor.

Standardizing the input data will support the effectiveness and stability of neural network structures. In this case, batch normalization is used for normalization; hence, this method also avoids excessive overfitting of the model. Pooling layer in the 1D-CNN structure can be viewed as down-sampling the convolutional layer to reduce the size of the parameters while conserving the vital features to make the next layer's determination faster. The dropout layer is an alternative to batch normalization besides the pooling layer when dealing with overfitting issues. In this layer, the disabled neurons are randomly removed from the training phase. Convolutional layers output a vector for input to a fully reconstructed layer, which is then flattened and reshaped. To complete the process, every label's probability is allocated to the final layer. This work used a 1D max-pooling technique to lessen the sampling of the convolution outputs by using two convolutional layers with two pooling layers. SoftMax, dropouts, and the full connection layer have been carefully crafted. A description of the target shape besides factors of all layers in the 1D-CNN model is given in Table 4.2.

Table 4.2 A description of each layer's parameters.

Layer (type)	Output shape	Number of parameters
Conv1D_1	(None, 5, 128)	768
Batch_normalization_1	(None, 5, 128)	512
Max_pooling1d_1	(None, 2, 128)	0
Conv1D_2	(None, 2, 64)	41,024
Batch_normalization_2	(None, 2, 64)	256
Max_pooling1d_2	(None, 1, 64)	0
Flatten	(None, 64)	0
Dense_1	(None, 64)	4160
Dropout_1	(None, 64)	0
Dense_2	(None, 32)	2080
Dropout_2	(None, 32)	0
Dense_3	(None, 6)	198
	Total parameters: 48,998	

4.6 Contact Elements for IoT and ML-based Data Visualization

The classification methods include DT, RF, XGBoost, and 1D-CNN in identifying fault conditions of the power equipment, including cyber-attacks [41]. Training and testing processes were generated from experimental data. Both model parameters and hyperparameters are types of parameters present in machine learning models. By analyzing the datasets, the models generate the model parameters automatically. Hyperparameters, like the learning rate parameter, can be adjusted to improve the performance of models. In order to construct machine learning models with the highest accuracy and the least error, the hyperparameter tuning process is required. The tuning process uses a variety of optimization algorithms, like grid search as well as random search. Grid search is a great way to overcome the limitations of manual search as it can provide a high level of hyperparameter optimization [42]. In order to discover an optimal set of parameters for RF and XGBoost models, the grid search algorithm is applied. A hyperparameter is optimized and well defined for each model following the tuning process. The extreme size of iterations with the XGBoost model was 400 for "*n* estimators," and the learning rate was 0.1, which allowed it to perform well while learning at a fast pace. Overfitting can be controlled by limiting the tree depth to 4. The "min child weight" value is equal to 5. The objective function "multi: softprob" can be used to optimize the learning process. As a result of optimizing the RF model, a maximum size of 800 iterations were chosen for "*n* estimators," an entropy criterion was used to minimalize the possibility of misclassifications, and log2 was selected as the number of features to be considered as a way of trying to determine the best split. Eqn (4.6) evaluates how well each machine learning technique performs.

$$\text{accuracy} = \frac{\text{TP} + \text{TN}}{\text{TP} + \text{FP} + \text{TN} + \text{FN}} \tag{4.6}$$

where TP denotes true positive, TN means true negative, FP depicts false positive, and, finally, FN represents false negative.

According to the following test scenarios, the proposed RF classifier model after being trained and tested is encoded with the IoT construction for categorizing the status of the motor online and presenting it with the IoT dashboard. The flowchart presented in Figure 4.12 shows the entire process of acquiring, validating, and visualizing data. Additionally, the next pseudo-code described in Algorithm 1 clarifies the procedure for the identification of a cyber-attack using the proposed IoT-based construction as well as the machine learning method.

Algorithm 1. The steps of attacks and faults recognition

Collect vibration data from the sensor
Send data to the cloud
Input data to machine learning model
Classify the data by machine learning model
Connect to IoT server
if the class of machine learning model = 0
Publish that the sensor data n is "Normal" and no attack
else if the class of machine learning model = 1
Publish that the machine has "Inner ring fault" and no attack
else if the class of machine learning model = 2
Publish that the machine has "Outer ring fault" and no attack
else if the class of machine learning model = 3
Publish that the data is "Fake data" and there is "Cyber-attacks"
End

4.6.1 Gas-insulated switchgear online monitoring

In this scenario, the insulating system represents both the normal and defective state of the developed IoT construction and the GIS insulation. A normal or healthy state is one in which there are no faults in the IoT architecture's insulation and there is no cyber-attack on the internet active network. The dashboard of the IoT platform in Figure 4.13 demonstrates the current GIS

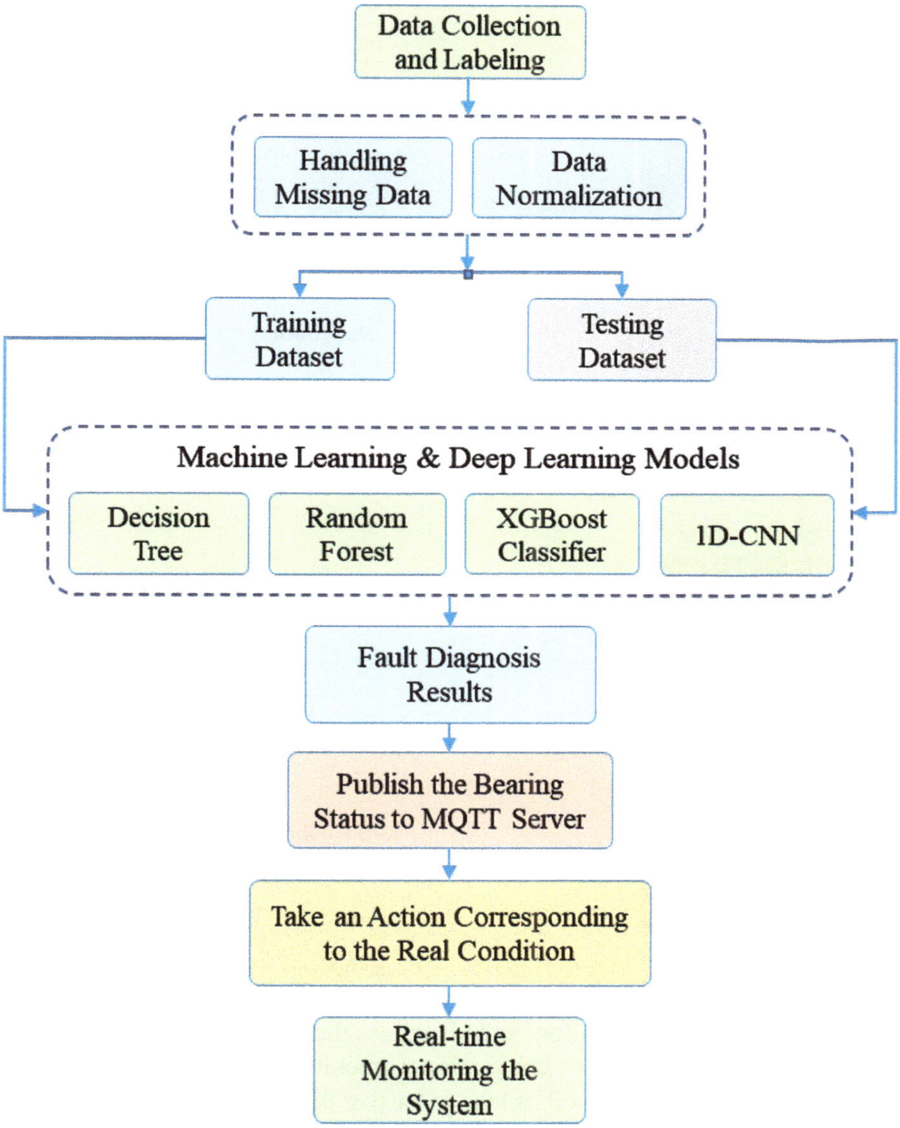

Figure 4.12 The proposed IoT platform flowchart paired with the numerous machine learning classifiers.

and its network status. In Figure 4.13, it is clear that the GIS is defect-free and the network is secure, which indicates no cyber-attacks are occurring and the IoT server is secure. Furthermore, the traffic light on the system is green, which indicates that the system functions properly. As a consequence, the

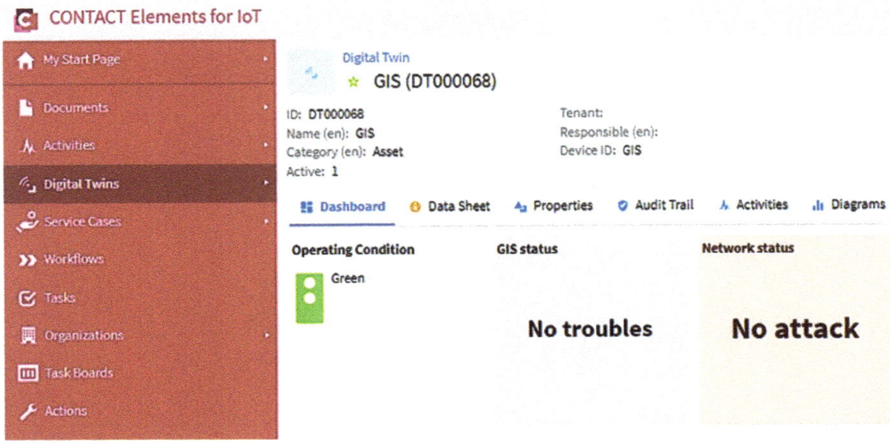

Figure 4.13 A normal GIS and network condition.

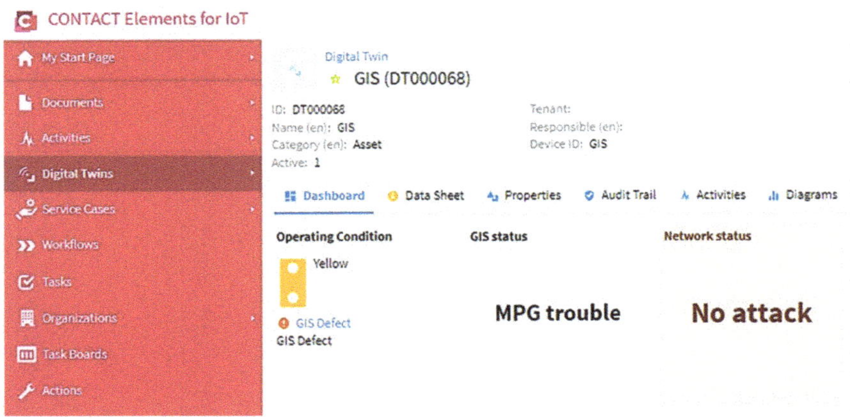

Figure 4.14 A defect in MPG's GIS and network condition.

developed IoT system monitors and visualizes the GIS in a clear, secure, and effective way as opposed to conventional tracking approaches that rely on manual measurements as well as take extensive time and cost a lot of money.

In Figure 4.14, the insulation status of the GIS is shown in an MPG defect scenario. The network is still secure as depicted in Figure 4.14, while there is an MPG defect in the system. Additionally, a yellow traffic light is presented to the user as an automatic warning about the defect state on the GIS for the purpose of maintaining it. It has been confirmed that the suggested machine learning methods work well with the IoT construction and that they can identify and visualize an MPG defect. Figure 4.15 displays the

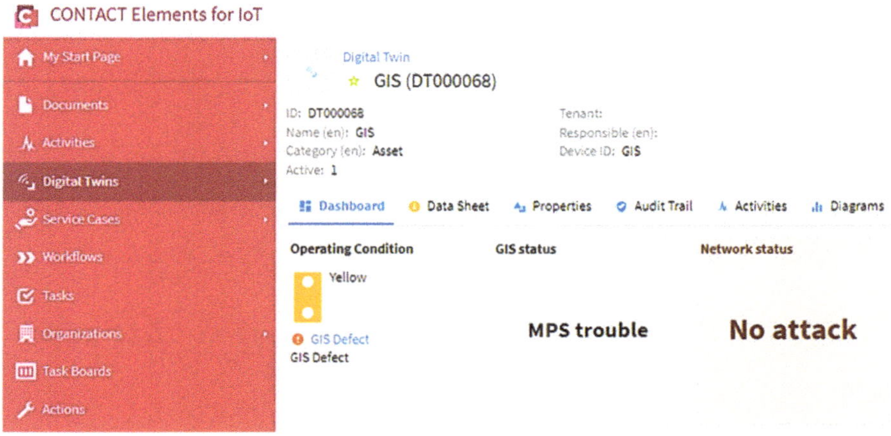

Figure 4.15 A defect in MPS's GIS and network condition.

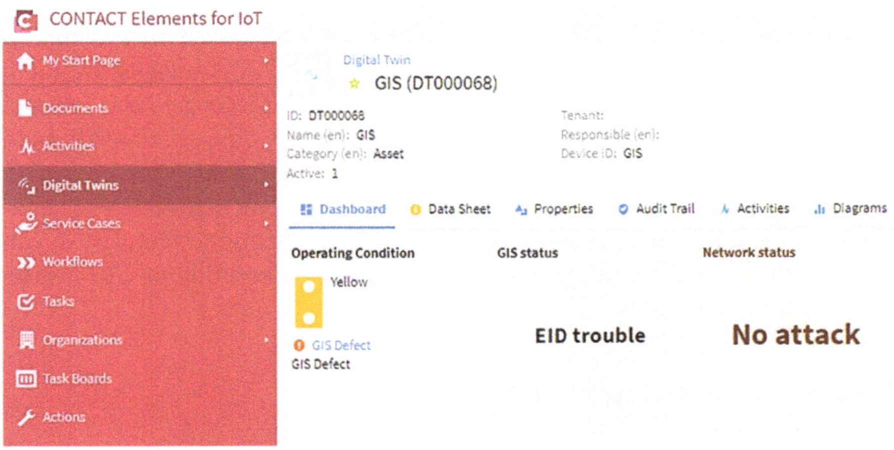

Figure 4.16 A defect in EID's GIS and network condition.

IoT platform's MPS fault and its network condition. It is apparent from the IoT dashboard, as seen in Figure 4.15, that the GIS insulation has an MPS fault, which is indicated by the traffic light changing to yellow. Moreover, the network is secure, and the transmission of data is secure; so the proposed machine learning technique can make better decisions about defect classification. Figure 4.16 clearly shows an EID defect in the GIS insulation. A warning alert is created by the IoT platform, and the light of the traffic indicator is altered to yellow to notify the user of the abnormal condition. Moreover, the network condition is still secure, which affirms the adequacy of the data

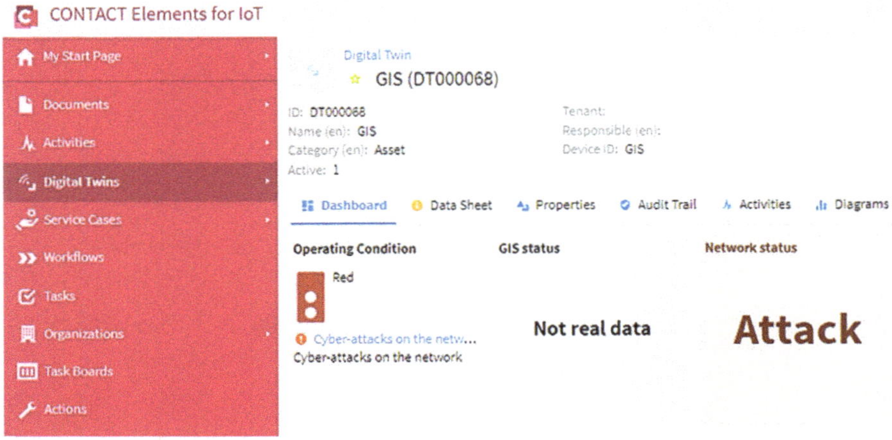

Figure 4.17 A scenario of cyber-attacks on GIS and the network.

transported to the GIS. Figure 4.17 illustrates that the internet network is insecure, meaning the IoT systems are vulnerable to cyber-attacks. As such, the GIS data has been forged. To maintain the IoT server and the internet network, the suggested IoT platform altered the traffic indicator to red light to alert the user about any abnormal attacks. In this test, the suggested machine learning method as well as the IoT construction are demonstrated to be capable of recognizing cyber-attacks besides informing the user effectively. It can be summarized that the suggested IoT can track the condition of GIS insulation much more reliably.

4.6.2 Power transformer online monitoring

The transformer status has been shown in both its normal and faulted states in this scenario. For instance, oil in a transformer in a normal state will have no dissolved gases or its concentration will be negligible. Consequently, there will be no electrical and thermal stresses on the oil transformer with high impact. The dashboard presented for the proposed IoT platform in Figure 4.18 demonstrates the concentration of dissolved gases besides the status of the oil transformer. In short, the IoT platform introduced here is capable of conveying the normal condition of a transformer clearly. Moreover, the traffic display's light color is green, which indicates that the transformer is in good condition.

Partial discharges on transformers are highly dangerous defects that must be detected early. Testing was performed to confirm the ability of the

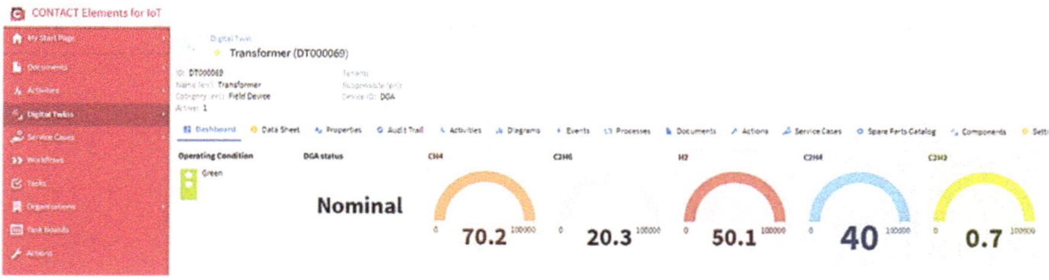

Figure 4.18 IoT dashboard showing normal transformer condition.

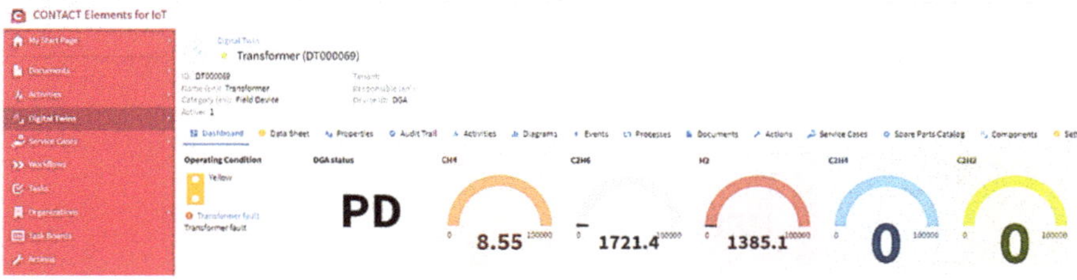

Figure 4.19 An IoT dashboard showing the transformer's condition of a PD fault.

IoT construction to detect PD and alert the user of transformer abnormalities using deep learning. In case of an abnormal condition, Figure 4.19 displays the concentrations of dissolved gases in the transformer and the transformer condition. As demonstrated in Figure 4.19, the introduced IoT platform dashboard displays the PD on the transformer properly. Similarly, traffic light color indicators are changed to red color to alert the operator of any abnormal states of the transformer, enhancing decision-making and keeping the transformer working healthy for a long period of time.

To keep the transformer working properly, a low-energy discharge must be detected in the transformer oil. The main purpose of this test is to show the capability of the IoT structure that advances deep learning to comprehend and visualize LEDs in transformer oil using the introduced IoT platform. Figure 4.20 displays the concentration of dissolved gases as well as the transformer status of the introduced IoT platform for LEDs. With the IoT structure, it is obvious that the LED defect can be presented effectively to the operator by creating an event and changing the traffic indicator to red color and informing the worker about the defect.

It is important to keep the LEDs in a healthy state by detecting high-energy discharge early. In order to determine whether the now presented deep-learning-based IoT structure is capable of distinguishing the HED from the oil in transformers, this test has been performed. By launching the IoT

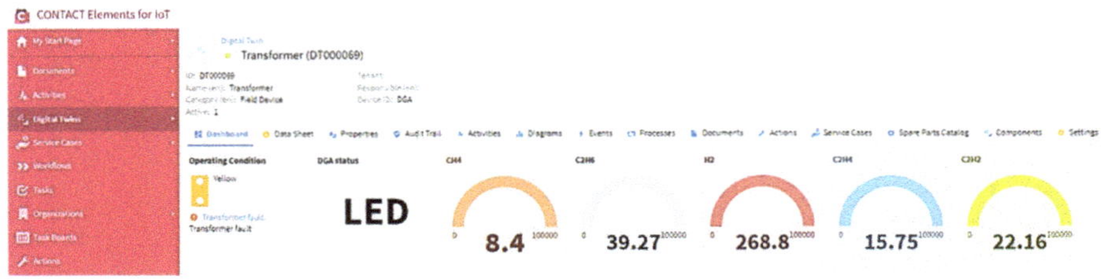

Figure 4.20 The IoT dashboard that displays the condition of the LED defect of a transformer owing to dissolved gases.

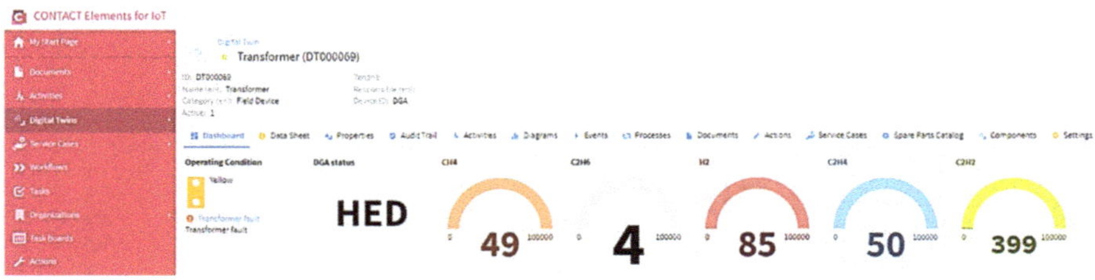

Figure 4.21 IoT dashboard showing the transformer condition of the HED failure.

Figure 4.22 The IoT dashboard that displays the transformer condition of the LT case.

dashboard, Figure 4.21 illustrates the status of the transformers as well as the concentrations of dissolved gases. Figure 4.21 illustrates how the new IoT structure combined with deep learning can detect and visualize high-energy discharge effectively.

Another sort of defect that affects the life of transformer oil is thermal faults, which must be detected so that the transformer maintains its efficiency. Thermal faults can be divided into three categories: low thermal (LT), medium thermal (MT), and high thermal (HT). The objective of this test is to demonstrate the capability of the IoT structure to notice and anticipate thermal faults in transformer oil using deep learning. Using the IoT platform dashboard, LT, followed by MT and HT faults, are displayed, respectively, as seen in Figures 4.22, 4.23, and 4.24. These figures demonstrate the

Figure 4.23 Transformer status in case of MT fault case on the IoT platform structure.

Figure 4.24 Transformer status on the IoT dashboard with the HT case.

Figure 4.25 IoT dashboard's internet network condition in the event of a cyber-attack.

best-in-class recognition of thermal defects in oil in the transformers by the IoT platform implemented with deep learning. To aid the operator, the IoT structure creates an event and changes the traffic indicator light color to red in order to alert them to the thermal fault in order to optimize the transformer condition decision-making process.

Cyber-attacks pose a greater threat to IoT systems than their execution. In this context, this test is designed to validate the capability of the presented IoT architecture with deep learning for recognizing cyber-attacks to provide reliable and secure remote monitoring for power transformers. As cleared in the dashboard of the newly developed IoT platform in Figure 4.25, the suggested IoT platform with deep learning recognizes cyber-attacks properly. To alert the worker about a cyber-attack on the wide-ranging network, the

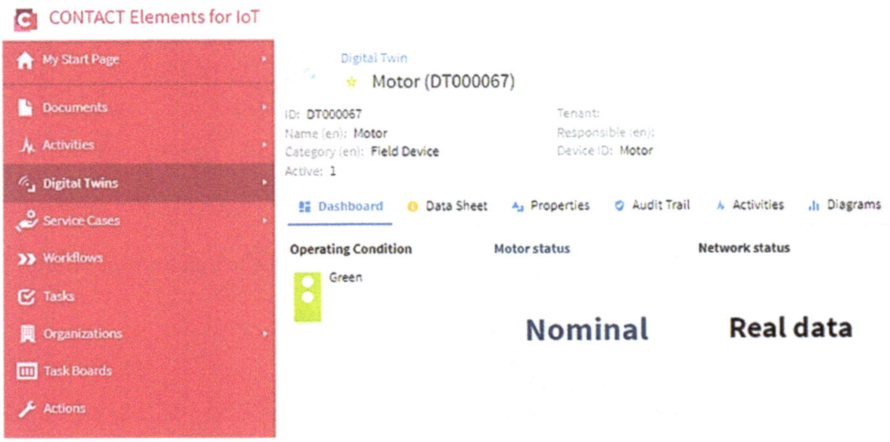

Figure 4.26 IoT dashboard showing motor status during normal operation.

proposed IoT systems change the traffic color to red to notify the event. In addition to the proposed procedure for detecting defects and cyber-attacks, the suggested IoT platform does an analysis of the latency and overhead associated with it. This IoT-based infrastructure with deep learning generally identifies transformer failures as well as cyber-attacks in a timely manner with minimal latency.

4.6.3 Online monitoring for induction machine

In the first scenario of the normal state as well as a stable network, online monitoring of the motor bearing is demonstrated [43]. Figure 4.26 displays the current bearing position and reliability of the internet network connection that is presented on the CONTACT Elements dashboard for an IoT structure. The motor status in the figure is normal, which means it is working properly and the bearing is not damaged. In addition, the light on the traffic alert is green, which indicates that the system is performing well. Further, Figure 4.26 demonstrates that the internet network is still secure and data transmission is actual. Testing the IoT architecture validates its reliability and facilitates the decision-making process regarding motor condition.

Using the IoT architecture and the developed machine learning algorithm, the capability of detecting bearing inner ring faults has been demonstrated. A bearing showing the condition of a stable network and an inner ring defect is shown in Figure 4.27. The IoT dashboard makes it clear that the bearing has an inner ring fault. It also confirms that the data transmitted via

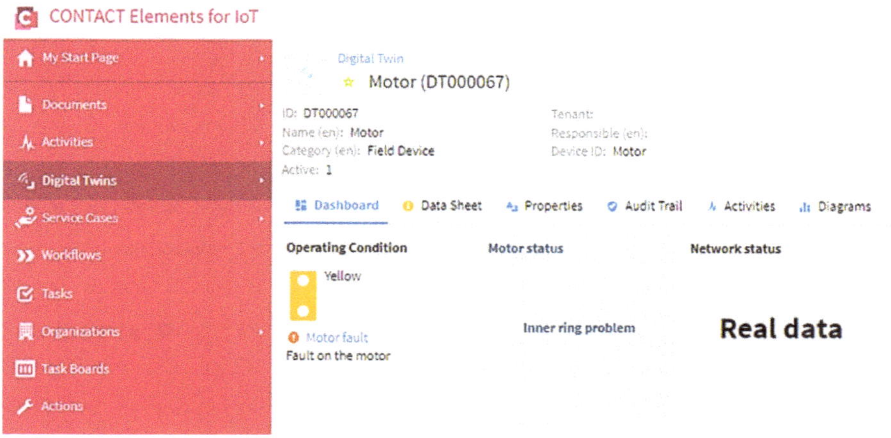

Figure 4.27 IoT dashboard that displays the motor condition in the case of the inner ring defect.

IoT is real. In addition, the traffic light turned yellow, which indicates a problem with the motor. The test results demonstrate that the suggested machine learning method, as well as the IoT architecture, can identify and visualize the fault in the inner ring of a motor bearing exclusive of any errors. The proposed methods also grow the reliability of the IoT paradigm by checking transmitted data for any cyber-attacks.

To demonstrate the effectiveness of the machine learning technique with IoT construction for detecting all fault cases in a motor bearing, an outer ring fault scenario is created. Figure 4.28 shows the bearing and network condition in the presence of a failed outer ring and safe data transfer. The IoT dashboard clearly shows that the machine learning procedure can identify a motor fault in the outer ring. In addition, the traffic indictor turned from the green light, which represents the normal state, into the yellow indictor to alert the worker of the abnormal condition caused by the outer ring defect of the motor bearing. On top of that, the dashboard displays that the network is still secure and that the transmitted data is actual.

The biggest challenge of implementing the IoT system is cybersecurity threats. Therefore, the implementation of this scenario is to establish the capability of the developed system in detecting cyber-attacks on the internet network. Cyber-attacks on the internet network will result in the transmitted data status being altered in Figure 4.29. From this figure, it is clear that the suggested machine learning algorithm can identify cyber-attacks on internet networks and that the proposed IoT construction is able to detect fake

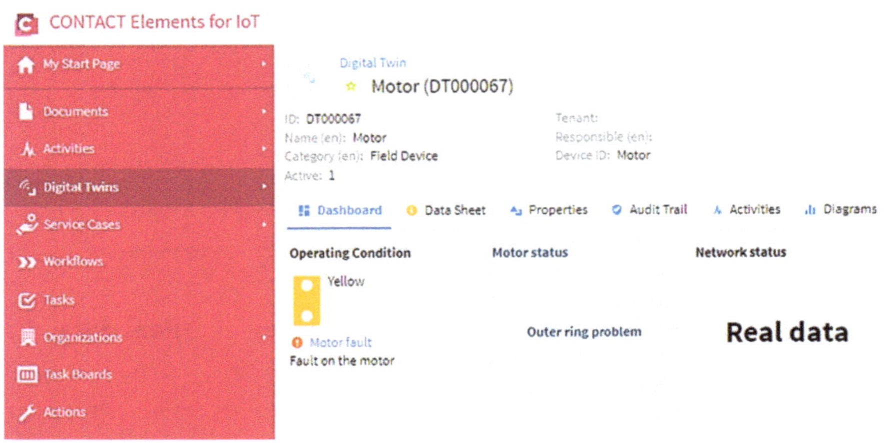

Figure 4.28 IoT dashboard that displays the motor status in the case of the outer ring defect.

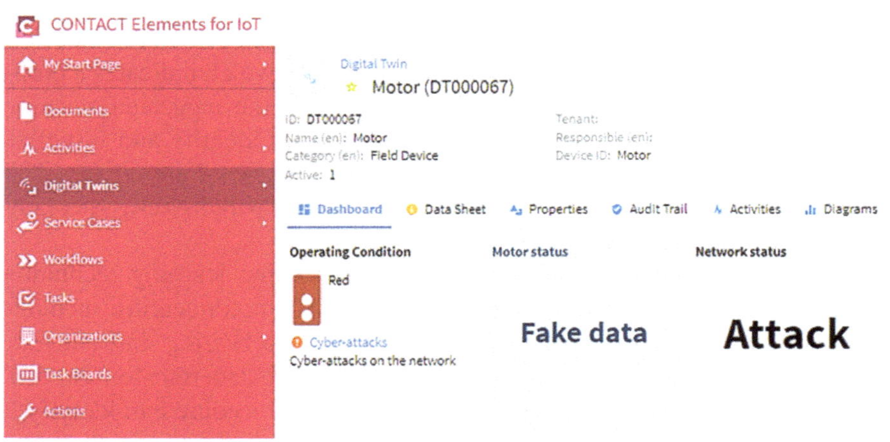

Figure 4.29 IoT dashboard showing the condition of the network in case of cyber-attacks.

information relating to motor status transmitted by the internet. A red light was added to the traffic light to signal that a cyber-attack was taking place on the network, which enhances decision-making and enables the network to be maintained and stabilized. Based on this test, machine learning using IoT construction can be used to deliver reliable motor status monitoring.

Acknowledgment

The authors would like to express their gratitude to the Contact Software company for supporting this work.

References

[1] N. Benzaoui, M.S. Gonzalez, M.V. Rivera, J.M. Estarán, H. Mardoyan, W. Lautenschlaeger, ... & S. Bigo, "DDN: Deterministic dynamic networks. In 2018 European Conference on Optical Communication (ECOC) (pp. 1–3) IEEE," September 2018. Rome, Italy.

[2] S. Redana, Ö. Bulakci, A. Zafeiropoulos, A. Gavras, A. Tzanakaki, A. Albanese, ... & Bernardos, "5G PPP Architecture Working Group: View on 5G Architecture," C. J., 2019,

[3] IoT Platform for Digital Business Models|CONTACT Software. Available online: https://www.contact-software.com/en/products/iot-platform-for-digital-business-models/?fbclid=IwAR0oYDd4qHpCd0BEZaGrL-HEAQGYoQ2BhBmDzbF35-cyM6QrNHAkziWDC8yo (accessed on May 6th, 2022).

[4] D.E.A. Mansour, K. Nishizawa, H. Kojima, N. Hayakawa, F. Endo, and H. Okubo, "Charge accumulation effects on time transition of partial discharge activity at GIS spacer defects," IEEE Trans. Dielectr. Electr. Insul., vol. 17, no. 1, pp. 247–255, Feb. 2010.

[5] D.E.A. Mansour, H. Kojima, N. Hayakawa, F. Endo, and H. Okubo, "Partial discharge detection at delamination of electrode/epoxy interface in GIS spacers," in Annual Report - Conference on Electrical Insulation and Dielectric Phenomena, CEIDP, 2009, pp. 11–14.

[6] D.E.A. Mansour, H. Kojima, N. Hayakawa, M. Hanai, and H. Okubo, "Physical mechanisms of partial discharges at nitrogen filled delamination in epoxy cast resin power apparatus," IEEE Trans. Dielectr. Electr. Insul., vol. 20, no. 2, pp. 454–461, 2013.

[7] D.E.A. Mansour, H. Kojima, N. Hayakawa, F. Endo, and H. Okubo, "Partial discharges and associated mechanisms for micro gap delamination at epoxy spacer in GIS," IEEE Trans. Dielectr. Electr. Insul., vol. 17, no. 3, pp. 855–861, Jun. 2010.

[8] D.A. Mansour, I.B.M. Taha, R. A. Farade, and N. I. B. Abdul Wahab, "Partial Discharge Diagnosis in GIS based on Pulse Sequence Features and Optimized Machine Learning Classification Techniques", Electric Power Systems Research, vol. 211, p. 108162, 2022.

[9] P. Wenger, M. Beltle, S. Tenbohlen, U. Riechert and G. Behrmann, "Combined Characterization of Free-Moving Particles in HVDC-GIS Using UHF PD, High-Speed Imaging, and Pulse-Sequence Analysis," IEEE Transactions on Power Delivery, vol. 34, no. 4, pp. 1540–1548, Aug. 2019.

[10] N. Abu Bakar and A. Abu-Siada, "A new method to detect dissolved gases in transformer oil using NIR-IR spectroscopy," IEEE Transactions

on Dielectrics and Electrical Insulation, vol. 24, no. 1, pp. 409–419, 2017.

[11] K. Chen et al., "All-optical photoacoustic multigas analyzer using digital fiber-optic acoustic detector," IEEE Transactions on Instrumentation and Measurement, vol. 69, no. 10, pp. 8486–8493, 2020.

[12] T. Yang, W. Chen and P. Wang, "A review of all-optical photoacoustic spectroscopy as a gas sensing method," Applied Spectroscopy Reviews, vol. 56no. 2, pp. 143–170, 2021.

[13] A. M. Alshehawy, D. A. Mansour, M. Ghali, M. Lehtonen, and M. M. F. Darwish, "Photoluminescence Spectroscopy Measurements for Effective Condition Assessment of Transformer Insulating Oil," Processes, vol. 9, no. 5, p. 732, Apr. 2021.

[14] F. Ma, Z. Liao, Y. Zhao, Z. Qiu, L. Wan, K. Li, G. Zhang, "Detection of trace C_2H_2 in N_2 buffer gas with cantilever-enhanced photoacoustic spectrometer," Optik, vol. 232, p. 166525, 2021.

[15] M. Badawi et al., "Reliable Estimation for Health Index of Transformer Oil Based on Novel Combined Predictive Maintenance Techniques," IEEE Access, vol. 10, pp. 25954–25972, 2022.

[16] A. Chatterjee, P. Bhattacharjee, N. K. Roy, and P. Kumbhakar, "Usage of nanotechnology-based gas sensor for health assessment and maintenance of transformers by DGA method," International Journal of Electrical Power & Energy Systems, vol. 45, no. 1, pp. 137–141, 2013.

[17] A.S.M. Iftekhar Uddin, U. Yaqoob, G.-S. Chung, "Dissolved hydrogen gas analysis in transformer oil using Pd catalyst decorated on ZnO nanorod array," Sensors and Actuators B: Chemical, vol. 226, pp. 90–95, 2016.

[18] J. Fan, F. Wang, Q. Sun, F. Bin, H. Ye and Y. Liu, "An online monitoring system for oil immersed power transformer based on SnO2 GC detector with a new quantification approach," IEEE Sensors Journal, vol. 17, no. 20, pp. 6662–6671, 2017.

[19] H. Cui, X. Zhang, G. Zhang, J. Tang, "Pd-doped MoS_2 monolayer: A promising candidate for DGA in transformer oil based on DFT method," Applied Surface Science, vol. 470, pp. 1035–1042, 2019.

[20] J. Fan, Y. Lu, J. Ding, A. Meng, Z. Tang, and J. Ye, "SOFC detector with OCA approach to quantify trace gases dissolved in transformer oil," IEEE Sensors Journal, vol. 20, no. 2, pp. 648–655, 2020.

[21] M. Elsisi, M.-Q. Tran, K. Mahmoud, D.-E. A. Mansour, M. Lehtonen, and M. M. F. Darwish, "Effective IoT-based deep learning platform for online fault diagnosis of power transformers against cyberattack and data uncertainties," Measurement, vol. 190, Art. no. 110686, 2022.

[22] M.-K. Liu and P.-Y. Weng, "Fault Diagnosis of Ball Bearing Elements: A Generic Procedure based on Time-Frequency Analysis," Measurement Science Review, vol. 19, pp. 185–194, August 2019.

[23] M. Elsisi, M.-Q. Tran, K. Mahmoud, M. Lehtonen, and M.M.F. Darwish, "Deep Learning-Based Industry 4.0 and Internet of Things towards Effective Energy Management for Smart Buildings," Sensors, vol. 21, no. 4, p. 1038, Feb. 2021.

[24] M. Elsisi, M. -Q. Tran, K. Mahmoud, D. -E. A. Mansour, M. Lehtonen and M. M. F. Darwish, "Towards Secured Online Monitoring for Digitalized GIS Against Cyber-Attacks Based on IoT and Machine Learning," IEEE Access, vol. 9, pp. 78415–78427, 2021.

[25] M. Elsisi, K. Mahmoud, M. Lehtonen, and M. M. F. Darwish, "Reliable Industry 4.0 Based on Machine Learning and IoT for Analyzing, Monitoring, and Securing Smart Meters," Sensors, vol. 21, no. 2, p. 487, Jan. 2021.

[26] M. Elsisi, M. -Q. Tran, K. Mahmoud, M. Lehtonen and M.M.F. Darwish, "Robust Design of ANFIS-Based Blade Pitch Controller for Wind Energy Conversion Systems Against Wind Speed Fluctuations," IEEE Access, vol. 9, pp. 37894–37904, 2021.

[27] J.G. Ferreira and A. Warzecha, "An application of machine learning approach to fault detection of a synchronous machine," in 2017 International Symposium on Electrical Machines (SME), 2017, pp. 1–6. Naleczow, Poland.

[28] M.N. Ali, K. Mahmoud, M. Lehtonen and M.M.F. Darwish, "An Efficient Fuzzy-Logic Based Variable-Step Incremental Conductance MPPT Method for Grid-Connected PV Systems," IEEE Access, vol. 9, pp. 26420–26430, 2021.

[29] M. Elsisi, K. Mahmoud, M. Lehtonen and M.M.F. Darwish, "An Improved Neural Network Algorithm to Efficiently Track Various Trajectories of Robot Manipulator Arms," IEEE Access, vol. 9, pp. 11911–11920, 2021.

[30] M. Elsisi, K. Mahmoud, M. Lehtonen and M. M. F. Darwish, "Effective Nonlinear Model Predictive Control Scheme Tuned by Improved NN for Robotic Manipulators," IEEE Access, vol. 9, pp. 64278–64290, 2021.

[31] D. Zhang, L. Qian, B. Mao, C. Huang, B. Huang, and Y. Si, "A Data-Driven Design for Fault Detection of Wind Turbines Using Random Forests and XGboost," IEEE Access, vol. 6, pp. 21020–21031, 2018.

[32] Y. Zhao, L. Yang, B. Lehman, J. d. Palma, J. Mosesian, and R. Lyons, "Decision tree-based fault detection and classification in solar photovoltaic arrays," in 2012 Twenty-Seventh Annual IEEE Applied Power Electronics Conference and Exposition (APEC), pp. 93–99, 2012.

[33] R. Benkercha and S. Moulahoum, "Fault detection and diagnosis based on C4.5 decision tree algorithm for grid-connected PV system," Solar Energy, vol. 173, pp. 610–634, 2018.
[34] C. Aldrich and L. Auret, "Fault detection and diagnosis with random forest feature extraction and variable importance methods," IFAC Proceedings Volumes, vol. 43, no. 9, pp. 79–86, 2010.
[35] T. Chen and C. Guestrin, "XGBoost: A Scalable Tree Boosting System," presented at the Proceedings of the 22nd ACM SIGKDD International Conference on Knowledge Discovery and Data Mining, San Francisco, California, USA, 2016. Available: https://doi.org/10.1145/2939672.2939785.
[36] Z. Wu, X. Wang, and B. Jiang, "Fault Diagnosis for Wind Turbines Based on ReliefF and eXtreme Gradient Boosting," Applied Sciences, vol. 10, no. 9, 2020.
[37] J. H. Friedman, "Greedy Function Approximation: A Gradient Boosting Machine," Annals of Statistics, 29. 1189–1232, 2001.
[38] S. Kiranyaz, T. Ince, and M. Gabbouj, "Real-Time Patient-Specific ECG Classification by 1-D Convolutional Neural Networks," IEEE Transactions on Biomedical Engineering, vol. 63, no. 3, pp. 664–675, 2016.
[39] S. Kiranyaz, O. Avci, O. Abdeljaber, T. Ince, M. Gabbouj, and D. J. Inman, "1D convolutional neural networks and applications: A survey," Mechanical Systems and Signal Processing, vol. 151, p. 107398, 2021.
[40] S. Kwon, G. Park, Y. Jang, J. Cho, M. Chu, and B. Min, "Determination of oil well placement using convolutional neural network coupled with robust optimization under geological uncertainty," Journal of Petroleum Science and Engineering, vol. 201, p. 108118, 2021.
[41] M. -Q. Tran et al., "Reliable Deep Learning and IoT-Based Monitoring System for Secure Computer Numerical Control Machines Against Cyber-Attacks with Experimental Verification," IEEE Access, vol. 10, pp. 23186–23197, 2022.
[42] T. Yu, & H. Zhu, "Hyper-parameter optimization: A review of algorithms and applications," 2020. arXiv preprint arXiv:2003.05689.
[43] M. -Q. Tran, M. Elsisi, K. Mahmoud, M. -K. Liu, M. Lehtonen and M. M. F. Darwish, "Experimental Setup for Online Fault Diagnosis of Induction Machines via Promising IoT and Machine Learning: Towards Industry 4.0 Empowerment," IEEE Access, vol. 9, pp. 115429–115441, 2021.

5

Effective Energy Management Amid the 4th Industrial Revolution

Amr S. Zalhaf[1,2], R. Ramadan[1,2], Qi Huang[1,3], Yang Han[1], and Olusola Bamisile[1,3]

[1]School of Mechanical and Electrical Engineering, University of Electronic Science and Technology of China, China
[2]Department of Electrical Power and Machines Engineering, Faculty of Engineering, Tanta University, Egypt
[3]Sichuan Industrial Intelligent and Applications Lab, Chengdu University of Technology, China

Abstract

This chapter presents a comprehensive review of the applications of the IoT and digital twin technologies in energy management in electrical power systems. The chapter highlights how IoT and the digital twin technologies can enable effective energy management by optimizing, controlling, and monitoring energy consumption. For instance, in smart homes, IoT devices are monitoring and managing the consumption of energy, while in smart buildings, digital twin can be utilized to optimize energy usage and reduce costs. Similarly, smart grids can leverage the IoT and digital twin technologies for improved monitoring and control of energy distribution. Furthermore, the chapter outlines the challenges and opportunities associated with implementing digital twin and IoT technologies for effective management of energy. Some identified challenges include data privacy and security concerns, integration with legacy systems, and the need for skilled professionals. The chapter concludes by giving future recommendations for research in this area, emphasizing the need for interdisciplinary collaboration and developing new digital twin and IoT solutions to address these challenges.

5.1 Introduction

The rapid expansion of urban areas has created numerous difficulties concerning energy consumption control. Innovative solutions have emerged to tackle energy management issues on the demand side. Energy management refers to the monitoring, regulation, and optimization of energy use to decrease energy consumption and cost while enhancing energy efficiency and sustainability and preserving consumers' comfort [1, 2]. In various industries, such as building automation, transportation, and manufacturing, energy management systems (EMSs) have become crucial. These systems are increasingly popular in promoting energy efficiency in residential and commercial smart grids. They use several factors, including energy cost, weather conditions, load profiles, and consumer satisfaction, to function [3].

To ensure secure and cost-efficient energy usage for customers, a reliable appliance load monitoring (ALM) system is a crucial component of a powerful energy management scheme [4]. ALM aims to achieve precise energy sensing and information on energy usage [5, 6]. This enables automated EMSs to identify high energy-consuming appliances and implement energy-saving strategies, such as rescheduling high-power operations during off-peak hours. ALM can be classified into two techniques: intrusive load monitoring (ILM) and non-intrusive load monitoring (NILM) [7]. ILM utilizes distributive sensing, whereas NILM is based on single-point sensing at the user's residence. Although ILM yields more accurate results, it has practical issues, such as the cost of the measurement system. On the other hand, NILM's significant advantage is its low maintenance and installation costs, as it requires fewer components and less space [8, 9].

According to [10], several technologies are utilized to enhance energy management, including:

1. Smart metering: These digital devices measure and transmit energy consumption data in real time, providing a comprehensive understanding of energy usage patterns. Smart meters can help in more effective and cost-efficient energy management [11].
2. Building automation: This system employs sensors and other modern electronic devices to regulate the energy usage of buildings, including heating, lighting, ventilation, and air conditioning (HVAC). This can help to minimize the consumption and costs of energy and enhance comfort and productivity [12].
3. Renewable energy sources (RESs): RESs such as solar systems and wind turbines that are used to generate electricity, lowering reliance on fossil fuels and decreasing greenhouse gas emissions [13].

4. Energy storage systems (ESSs): These can store surplus energy produced from renewable sources for future use, enhancing energy security and reducing the requirement for traditional energy sources. ESS can store and deliver energy at a high response rate and efficiency, enabling them to alleviate the mismatch between power generation and demand without increasing power generation or causing energy waste. Integrating ESSs with RESs is critical in resolving the intermittency of power supply and significantly improving power supply availability, efficiency, and stability [14].

5. Internet of Things (IoT) and digital twin: IoT and digital twin technologies can collect and analyze data on energy consumption, detecting patterns and opportunities for energy savings. A digital twin is a virtual replica of a physical asset or system, like a building or a manufacturing facility, that can simulate the performance of the physical asset and identify potential issues [15].

6. Energy management software: This software provides a unified solution for monitoring and managing the consumption of energy across different buildings, manufacturing facilities, and cities [16].

A combination of different technologies can be used to improve energy management, leading to decreased energy consumption and costs. The selection of the most suitable technology will depend on various factors, such as the desired energy management goals, the existing infrastructure, and the resources available. The following sections will specifically discuss the utilization of IoT and digital twin technologies to enhance energy management, which involves collecting, analyzing, and simulating energy consumption data to identify patterns and opportunities for energy conservation.

5.2 Energy Management using IoT

As energy demand and expenses continue to rise, organizations seek innovative approaches to monitor, regulate, and conserve energy. Smart EMSs can reduce costs while satisfying energy demands. Advanced technologies like IoT and big data can assist in managing energy usage across commercial, industrial, and residential sectors. IoT technology is employed to monitor, regulate, and optimize energy consumption in various industries, including healthcare, retail, manufacturing, and energy. By connecting devices and sensors to the internet, IoT technology can collect and transmit data on energy usage, enabling data analysis to detect patterns and optimize energy consumption.

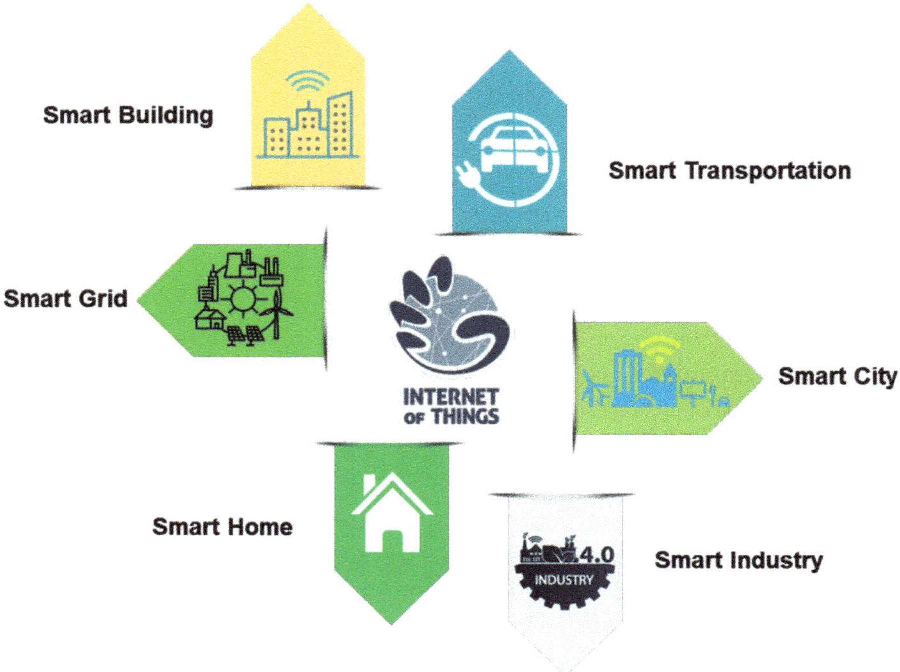

Figure 5.1 Structure of IoT solutions in energy management.

The following sections describe several ways in which IoT can be applied in energy management and power consumption. These include smart buildings, smart homes, smart grids, smart cities, smart transportation, and Industry 4.0. Figure 5.1 provides an illustration of these applications.

5.2.1 Energy management in smart homes using IoT

In this section, the focus is on the solutions for energy management in smart homes utilizing IoT. A study presented in [17] proposed a method for managing energy consumption of electrical appliances during peak periods using IoT. The researchers developed a monitoring device with Wi-Fi module, microcontroller, flow sensor, and power supply using the EasyEDA program. The device was utilized to monitor energy consumption of appliances over a one-day period and calculate consumption using smart processing and data collection. However, this method requires consumers to temporarily turn off certain appliances during peak consumption periods. The authors suggested a list of appliances to turn off and utilized guidelines from electricity distribution companies, including pricing information for various consumption levels, to help consumers adjust their usage patterns. The paper did not provide a specific algorithm but presented a graph comparing the price-to-energy ratio and evaluated the effectiveness of the device by comparing energy consumption with and without its use.

A method for implementing a self-learning home management system using machine learning in an IoT environment was presented in [18]. The system was designed to monitor electricity consumption and optimize homes by reducing unnecessary power usage and lowering costs. It included features such as power alerts, price predictions, and price clustering developed through machine learning technology. The system was simulated using a house, and smart plug-ins were installed to monitor power consumption. Data analysis was utilized to adjust power usage based on the homeowner's habits. Another study proposed a smart home management system that remotely monitors and controls home appliances using the IoT technology [19]. The system consisted of a central hub device connected to various home appliances such as lights, fans, and air conditioners via wireless communication. The user could access the central hub device through a mobile application to monitor and control the appliances. The system also allowed the user to set up rules and schedules for the appliances to automate their operation. The authors discussed the implementation details of the system, including the hardware and software components used, and presented the testing and evaluation results.

Moreover, the authors in [20] conducted a study to examine the factors that affect consumer perception of risk related to using IoT-based home energy management services. They identified technical, service, personal, and environmental factors that influence consumer risk perception. Technical factors, including security and reliability, were found to have the most significant impact on risk perception. Service factors, such as customer service and transparency, had a moderate impact, while personal and environmental factors had a relatively small impact. The study also highlighted key attributes of IoT-based home energy management services that can help reduce consumer risk perception and increase trust and adoption of these services. These findings can be useful in developing more secure and reliable IoT-based home energy management services that meet consumer needs and expectations.

Using machine learning algorithms, like LSTM, can help in developing a more accurate prediction model for energy consumption patterns in smart homes. The proposed system in [21] provided a flexible and scalable approach to adapt to different types of home appliances and changing energy consumption patterns. This can be beneficial in optimizing energy usage, reducing energy costs, and promoting sustainability. The system's ability to automatically collect data and generate insights can also provide homeowners with a better understanding of their energy consumption patterns, allowing them to make informed decisions about their energy usage. Overall, the integration of machine learning algorithms with IoT technology can significantly improve smart home EMSs, leading to more efficient and sustainable energy use.

5.2.2 Energy management in smart buildings using IoT

In industries such as building automation, utilizing IoT-connected devices like sensors and smart meters has become increasingly popular for regulating energy consumption within buildings. The utilization of IoT technology in smart buildings has been shown to enhance energy efficiency and reduce environmental impact. Research has indicated that effective communication between a building's systems can substantially reduce energy consumption. Therefore, the growth in networking, sensing, and computation technologies has made IoT a crucial factor in designing and managing intelligent structures within the built environment. For instance, sensors can be used to detect room occupancy in a smart building and adjust lighting and temperature settings accordingly, thereby minimizing energy wastage [22, 23]. Monitoring systems and smart meters can monitor energy consumption in real time and provide building managers and energy providers with data to optimize energy usage. This approach can help identify energy-saving trends and opportunities, leading to more sustainable and efficient energy consumption.

In a previous study [24], the authors developed a real-time building EMS that utilized IoT and data analytics to monitor energy usage and optimize energy usage on a university campus. Another study [25] proposed a smart building management system that leveraged IoT technology to enhance building operations and energy efficiency in Vietnam. The proposed system incorporated various sensors, gateways, and a centralized management platform to monitor and control different building systems. The authors also assessed the feasibility of implementing such a system in Vietnam from a business perspective, including the costs and benefits for building owners, facility managers, and tenants. The study found that the system had the potential to significantly reduce energy consumption, improve building operations, and enhance tenant satisfaction. The authors also determined that the system was economically viable, with benefits exceeding costs over the long term. In addition, a previous work [26] introduced an IoT energy platform (IoTEP) that aimed to tackle energy management issues, such as energy wastage and excessive consumption, in homes. The authors recognized that current methods, such as apartment sensing systems, were insufficient in addressing energy loss. Therefore, they proposed IoTEP to optimize energy management in buildings by collecting data, estimating energy savings, and providing personalized energy analysis for future developments.

5.2.3 Energy management in smart grids using IoT

Nowadays, IoT technology is increasingly employed in smart grid systems to optimize and enhance energy distribution, consumption, and management [27]. IoT devices, including sensors and smart meters, are strategically installed throughout the grid to gather data on energy consumption, generation, and distribution. This data is then analyzed to improve the grid's operations. By collecting data on energy usage and demand, grid operators can dynamically manage the flow of energy to meet changing demand, minimizing the risk of blackouts or brownouts. In addition, IoT can facilitate integrating RES into the grid, increasing its sustainability and decreasing its carbon footprint. IoT can also enhance outage management in smart grids. IoT sensors and devices can rapidly detect power outages, enabling grid operators to respond more efficiently and quickly to restore power. Furthermore, IoT can monitor the health and performance of the grid infrastructure, providing early warnings of potential issues, and enabling grid operators to take proactive measures to avoid or mitigate outages.

Pawar *et al.* [27] proposed a smart EMS in smart grids, which utilizes IoT technology for accurate prediction and load management of renewable energy generation. The system aims to overcome the challenges of managing and integrating RES into the grid. Real-time monitoring and control of energy consumption and generation are provided, and the authors claimed that their system offers improved accuracy in energy forecasting and optimized load management, resulting in greater efficiency and sustainability in the energy sector. In [28], the authors aimed to enhance energy management and demand reduction in a power distribution system by employing IoT and Fog technology. Their solution involved both consumers and utilities in managing energy consumption and reducing demand. The IoT collected data enabled the prediction of energy consumption and peak hours for the next three days, which could then be conveyed to users through a demand response program. This approach decreased energy consumption during peak hours, leading to lower energy costs for both residents and utilities. Consumers can allow utilities to manage and automatically control their devices to optimize energy usage.

5.2.4 Energy management in smart industry (Industry 4.0) using IoT

Industry 4.0 pertains to the Fourth Industrial Revolution, which encompasses incorporating advanced digital technologies into industrial processes and operations to improve efficiency, reduce costs, and enhance product quality. The main objective of Industry 4.0 is to establish intelligent and interconnected systems in manufacturing and production that can communicate and cooperate in real time, ultimately leading to a more flexible, efficient, and sustainable production environment that can quickly and effectively respond to market changes and customer requirements [29]. Within Industry 4.0, IoT-based devices and sensors can be employed to monitor and manage energy consumption in manufacturing facilities. By collecting data on the energy use of equipment and processes, sensors can assist in optimizing energy utilization and reducing downtime through data analysis.

A study in [30] was conducted to examine the use of IoT-enabled devices and sensors in Industry 4.0 to monitor energy consumption in manufacturing facilities and optimize energy usage. The authors reviewed the most common cybersecurity technologies in IoT and Industry 4.0 and analyzed their benefits for smart manufacturing while also highlighting the challenges of current cyber-attacks and offering some solutions. In addition, Kiran *et al.* [31] proposed a new power management scheme for the IEEE 802.15.4 MAC layer in IoT networks that aimed to minimize power usage while achieving customer-specified reliability at the nodes. The proposed scheme utilized 3D Markov chain and M/G/1/K queue algorithms to model the IEEE 802.15.4 MAC layer and form knots in queues. The study focused on the impact of limited storage capacity at the node level on the performance of the MAC layer, which is responsible for data flow control on a network. The goal of the study was to address the challenges of industrial IoT networks with limited energy and storage resources and the impact on the MAC layer's performance. The results indicated that the proposed scheme improved the performance of the MAC layer under limited storage constraints, making it a valuable contribution to the field of industrial IoT and energy management.

5.2.5 Energy management in the transportation industry using IoT

IoT technology is increasingly being utilized in the transportation industry to optimize various aspects such as fleet management, maintenance, traffic management, and passenger experience. By utilizing connected vehicles, accidents can be reduced, and traffic flow can be improved. Monitoring systems and IoT sensors can be used to diagnose maintenance requirements and track vehicle performance, leading to enhanced efficiency and reduced downtime. Furthermore, IoT can improve the passenger experience through the use of in-vehicle entertainment and information systems, smart ticketing, and real-time travel information. The use of IoT in the transportation sector has the potential to increase safety, efficiency, and customer satisfaction. For example, IoT-enabled devices can be used to manage the energy consumption of electric vehicles (EVs) and plug-in hybrid electric vehicles (PHEVs). Data collected by sensors and communication devices on the state of the battery, driving conditions, and energy consumption can be analyzed to improve the vehicle's energy consumption and overall efficiency. A study was conducted before [32] in which the authors reviewed current energy management strategies for EVs and PHEVs and discussed their potential for enhancing the energy efficiency of these vehicles.

5.2.6 Energy management in smart cities using IoT

Smart cities are utilizing the IoT technology to enhance their energy systems and minimize CO_2 emissions. One method involves using IoT sensors and devices to monitor energy consumption in real time, enabling city officials to pinpoint areas where energy usage can be optimized. Predictive analytics and machine learning algorithms can be applied to anticipate energy demand and improve energy distribution networks, decreasing energy waste and ensuring efficient energy production [33]. Furthermore, some smart cities have introduced microgrids, which are self-contained energy systems capable of functioning autonomously from the main grid. This permits cities to manage energy consumption more efficiently and offers a backup energy source in the event of an outage. The integration of IoT technology in smart cities can significantly increase energy efficiency, minimize waste, and create more sustainable and resilient energy systems. IoT-enabled devices and sensors can be implemented to supervise and regulate energy consumption in buildings, transportation systems, and infrastructure. Energy consumption and traffic patterns data can be collected and analyzed to optimize energy usage and minimize traffic congestion [34, 35]. Smart grids are able to improve energy management in smart cities by integrating RES, like wind and solar power, into the energy supply.

A review of IoT-based smart EMSs in smart cities and their potential for decreasing energy consumption and promoting renewable energy sources was discussed in [36, 37]. In addition, a system utilizing traceable graphical contexts was proposed in [38] for data management in scalable IoT systems within smart cities. The system used diagrams to depict its design and was implemented in an application that enables users to track context designs while also safeguarding data. The performance of the system was evaluated for error detection, but its context awareness framework was restricted and requires further development to integrate additional entities. In another work by Liu *et al.* [39], a model for energy efficiency in fog nodes of IoT environments was introduced, using deep reinforcement learning. The paper focused on energy management in smart cities and presented a software model of an IoT-based environment with edge computation. Reliable control measures are essential for the complexities of the smart city environment, and reinforcement learning is regarded as a promising technology to address these challenges.

In general, implementing IoT technology for energy management purposes can provide valuable information regarding energy consumption patterns and reveal opportunities for reducing energy usage, ultimately resulting in more efficient and sustainable energy utilization in different sectors.

5.3 Energy Management Using the Digital Twin Technology

The digital twin is a digital replica of a physical asset or system, such as a building or manufacturing facility. It can be used to simulate the performance of the physical asset and identify potential problems [40]. In the context of energy management, a digital twin can be used to model the energy consumption of a building or manufacturing facility and identify opportunities for energy savings. The digital twin technology can be used in various industries to improve energy management by simulating physical assets' performance and identifying potential energy consumption problems. It allows design, testing, and optimization solutions before they are implemented in the real world, reducing energy consumption and costs while also improving energy efficiency and sustainability [41]. An overview of the digital twin technology concept is illustrated in Figure 5.2.

The utilization of the digital twin technology in building automation enables building managers to simulate energy consumption and identify areas for energy savings. By creating a virtual copy of a building, managers can test various scenarios, such as changes in occupancy, lighting, and temperature, to assess their impact on energy usage. In a study by [42], the authors reviewed the present state of digital twin technology in energy management in the construction industry and its potential to optimize energy consumption in buildings. They also analyzed the current status and development history

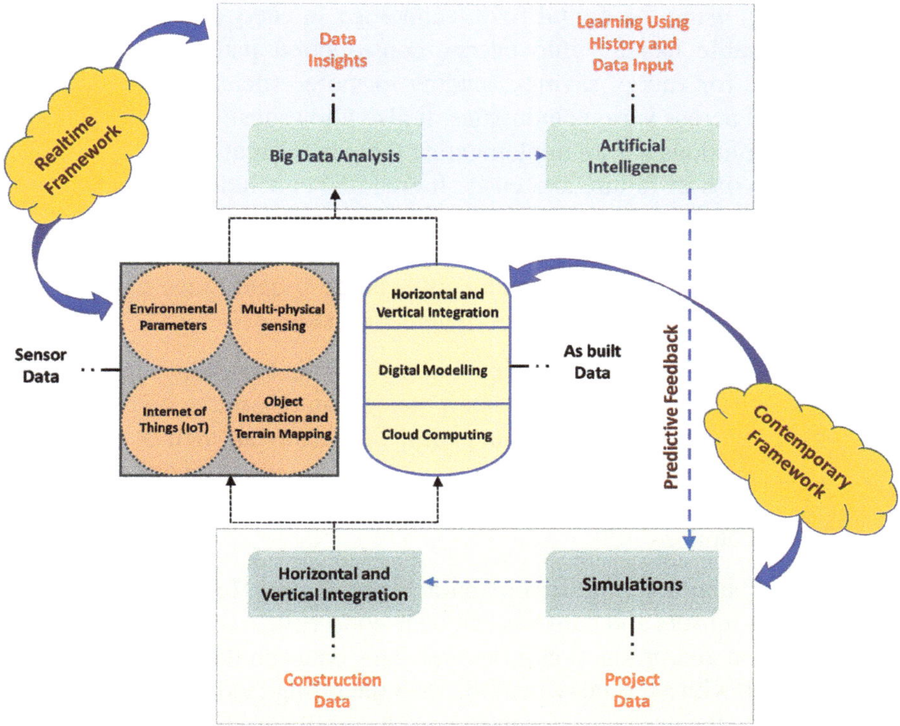

Figure 5.2 Schematic diagram of the digital twin technology [15].

of the digital twin concept, the essential technologies involved, and how the concept is utilized in various project stages, including building information modeling, system integrity, facility management, monitoring, logistics, and energy simulation.

In the manufacturing industry, the digital twin technology can be used to simulate the energy consumption of a manufacturing facility and identify opportunities for energy savings. It allows to test different configurations of the equipment and the process, evaluate the performance, and optimize the energy usage before making any changes in the real facility [43]. On the other hand, in smart cities, the digital twin technology can be used to simulate the energy consumption of a city and identify opportunities for energy savings. A digital twin of a city can be used to model the energy consumption of buildings, transportation systems, and infrastructure and evaluate the impact of different scenarios, such as changes in population density, the integration of renewable energy sources, and the implementation of smart grid systems [44].

Overall, using the digital twin technology in energy management can provide valuable insights into energy consumption patterns and identify opportunities for energy savings, leading to more efficient and sustainable energy usage across various industries. It also allows testing, validating, and optimizing solutions before implementing them in the real world, resulting in reducing the risk of failure and costs. It could be concluded that the combination of IoT and digital twin technologies can provide valuable insights into energy consumption patterns and identify opportunities for energy savings, leading to more efficient and sustainable energy usage. Some examples of how these technologies can be used together include:

1. Smart building energy management: In a smart building, IoT-enabled devices such as sensors and smart meters can be used to collect data on energy consumption and occupancy. This data can then be used with a digital twin of the building to simulate the energy consumption of the building under different scenarios and identify the most energy-efficient solutions [45].
2. Smart manufacturing: In a manufacturing facility, IoT-enabled devices such as sensors and cameras can be used to collect data on energy consumption and production processes. This data can then be used in conjunction with a digital twin of the manufacturing facility to simulate and optimize the energy consumption of the facility, reduce downtime, and improve overall efficiency [46].
3. Smart cities: In a smart city, IoT-enabled devices such as sensors and cameras can be used to collect data on energy consumption, traffic patterns, and other indicators. This data can then be used in conjunction with a digital twin of the city to simulate energy consumption [47].

5.4 Challenges and Future Recommendations

The implementation of the IoT and digital twin technologies for effective energy management presents both challenges and opportunities. Some of the challenges associated with the implementation of these technologies include [48]:

1. Data privacy and security concerns: As the IoT and digital twin technologies involve using a large amount of data, ensuring data privacy and security is critical. This can be challenging as cyber threats continue to evolve, and there is a need to develop new and robust security measures to protect sensitive data.

2. Integration with legacy systems: Many existing electrical power systems are based on legacy technologies, which may not be compatible with the IoT and digital twin technologies. This makes integration challenging, requiring significant investment and technical expertise to transition to the new system.

3. Skilled workforce: Implementing the IoT and digital twin technologies requires a highly skilled workforce that can design, develop, and maintain these systems. However, there is a shortage of professionals with the necessary expertise, which can hinder the implementation of these technologies.

On the other hand, the implementation of the IoT and digital twin technologies also presents numerous opportunities for effective energy management, including [49]:

1. Real-time monitoring and control: IoT and digital twin technologies can enable real-time monitoring and control of energy consumption, which can help optimize energy usage and reduce costs.

2. Improved decision-making: IoT and digital twin technologies can provide valuable insights into energy consumption patterns, which can help decision-makers develop more effective energy management strategies.

3. Sustainability: IoT and digital twin technologies can help promote sustainable energy practices by enabling the integration of renewable energy sources and reducing carbon emissions.

To conclude this chapter, it is important to highlight the need for future research in the field of IoT and digital twin technologies for effective energy management. Some recommendations can be summarized as follows:

1. Interdisciplinary collaboration: The implementation of IoT and digital twin technologies for effective energy management requires cooperation between various disciplines, including engineering, computer science, and energy management. Future research should prioritize interdisciplinary collaboration to develop solutions that address the challenges associated with these technologies.

2. Development of new IoT and digital twin solutions: The development of new and innovative IoT and digital twin solutions that can address the challenges associated with these technologies is essential. This includes the development of robust security measures, integration with legacy systems, and the creation of new algorithms for real-time monitoring and control of energy consumption.

3. Evaluation of the social and economic impact: Future research should focus on evaluating the social and economic impact of implementing IoT and digital twin technologies for effective energy management. This includes assessing the impact on energy consumption patterns, carbon emissions, and the cost-effectiveness of these technologies.

4. Training and education: The implementation of IoT and digital twin technologies requires a highly skilled workforce. Therefore, future research should focus on developing training and education programs to address the skills gap and equip professionals with the necessary expertise to implement these technologies effectively.

5.5 Conclusion

In summary, IoT and digital twin are perfect examples of how the energy management industry is transitioning from a siloed approach to a holistic one. This chapter provides an overview of the applications of IoT and digital twin technologies in the context of electrical power systems for effective energy management. The chapter also examines the use of IoT and digital twin technologies for energy management in various settings, such as smart homes, smart buildings, smart grids, smart industry, smart transportation, and smart cities. By integrating IoT, digital twin technologies, and advanced analytics (cloud services), organizations can empower consumers to become true partners in the demand-side management of their energy consumption. This will ultimately lead to more efficient and sustainable energy usage, reduced energy costs, and improved energy security. More research is needed to explore the full potential of these technologies and identify the most effective ways of implementing them.

Moreover, the challenges and opportunities associated with implementing IoT and digital twin technologies are highlighted for effective energy management, including data privacy and security concerns, integration with legacy systems, the need for a highly skilled workforce, real-time monitoring and control, improved decision-making, and sustainability. Finally, the chapter provides recommendations for future research, emphasizing the need for interdisciplinary collaboration, developing new IoT and digital twin solutions, evaluating social and economic impact, and training and education programs.

Acknowledgment

This work was supported by Sichuan Provincial Postdoctoral Science Foundation under grant No. 246861.

References

[1] Al-Ali, A.R., Zualkernan, I.A., Rashid, M.; Gupta, R., Alikarar, M. A Smart Home Energy Management System Using IoT and Big Data Analytics Approach. *IEEE Transactions on Consumer Electronics* 2017, *63*, 426–434, doi:10.1109/TCE.2017.015014.

[2] Medojevic, M., Díaz Villar, P., Cosic, I., Rikalovic, A., Sremcev, N., Lazarevic, M. Energy Management in Industry 4.0 Ecosystem: A Review on Possibilities and Concerns. In Proceedings of the Annals of DAAAM and Proceedings of the International DAAAM Symposium, Danube Adria Association for Automation and Manufacturing, DAAAM, 2018, Vol. 29, pp. 0674–0680.

[3] Yaïci, W., Krishnamurthy, K., Entchev, E., Longo, M. Recent Advances in Internet of Things (IoT) Infrastructures for Building Energy Systems: A Review. *Sensors* 2021, *21*, 1–40.

[4] Hasan, M.M., Chowdhury, D., Khan, M.Z.R. Non-Intrusive Load Monitoring Using Current Shapelets. *Applied Sciences (Switzerland)* 2019, *9*, 1–22, doi:10.3390/app9245363.

[5] Saeed, S., Agbossou, K., Kelouwani, S., Cardenas, A. Non-Intrusive Load Monitoring through Home Energy Management Systems: A Comprehensive Review. *Renewable and Sustainable Energy Reviews* 2017, *79*, 1266–1274, doi:10.1016/j.rser.2017.05.096.

[6] Meehan, P., McArdle, C., Daniels, S. An Efficient, Scalable Time-Frequency Method for Tracking Energy Usage of Domestic Appliances Using a Two-Step Classification Algorithm. *Energies (Basel)* 2014, *7*, 7041–7066, doi:10.3390/en7117041.

[7] Fan, W., Liu, Q., Ahmadpour, A., Gholami Farkoush, S. Multi-Objective Non-Intrusive Load Disaggregation Based on Appliances Characteristics in Smart Homes. *Energy Reports* 2021, *7*, 4445–4459, doi:10.1016/j.egyr.2021.07.033.

[8] Aladesanmi, E.J., Folly, K.A. Overview of Non-Intrusive Load Monitoring and Identification Techniques. *IFAC-Papers OnLine* 2015, *48*, 415–420, doi:10.1016/j.ifacol.2015.12.414.

[9] Zoha, A., Gluhak, A., Imran, M.A., Rajasegarar, S. Non-Intrusive Load Monitoring Approaches for Disaggregated Energy Sensing: A Survey. *Sensors* 2012, *12*, 16838–16866, doi:10.3390/s121216838.

[10] Roosa, S.A., Doty, S., Turner, W.C. *Energy Management Handbook*, 2020,

[11] Pitì, A., Verticale, G., Rottondi, C., Capone, A., lo Schiavo, L. The Role of Smart Meters in Enabling Real-Time Energy Services for Households: The Italian Case. *Energies (Basel)* 2017, *10*, doi:10.3390/en10020199.

[12] Belussi, L., Barozzi, B., Bellazzi, A., Danza, L., Devitofrancesco, A., Fanciulli, C., Ghellere, M., Guazzi, G., Meroni, I., Salamone, F., et al. A Review of Performance of Zero Energy Buildings and Energy Efficiency Solutions. *Journal of Building Engineering* 2019, *25*, 100772, doi:10.1016/J.JOBE.2019.100772.
[13] Iris, Ç., Lam, J.S.L. A Review of Energy Efficiency in Ports: Operational Strategies, Technologies and Energy Management Systems. *Renewable and Sustainable Energy Reviews* 2019, *112*, 170–182, doi:10.1016/J.RSER.2019.04.069.
[14] Murty, V.V.S.N., Kumar, A. Multi-Objective Energy Management in Microgrids with Hybrid Energy Sources and Battery Energy Storage Systems. *Protection and Control of Modern Power Systems* 2020, *5*, doi:10.1186/s41601-019-0147-z.
[15] Bhatti, G., Mohan, H., Raja Singh, R. Towards the Future of Smart Electric Vehicles: Digital Twin Technology. *Renewable and Sustainable Energy Reviews* 2021, *141*, 110801, doi:10.1016/J.RSER.2021.110801.
[16] Khan, M.W., Wang, J., Ma, M., Xiong, L., Li, P., Wu, F. Optimal Energy Management and Control Aspects of Distributed Microgrid Using Multi-Agent Systems. *Sustain Cities Soc* 2019, *44*, 855–870, doi:10.1016/J.SCS.2018.11.009.
[17] Al-Oudat, N., Aljaafreh, A., Saleh, M., Alaqtash, M. Iot-Based Home and Community Energy Management System in Jordan. In Proceedings of the Procedia Computer Science, 2019, Vol. 160.
[18] Li, W., Logenthiran, T., Phan, V.T., Woo, W.L. Implemented IoT-Based Self-Learning Home Management System (SHMS) for Singapore. *IEEE Internet Things J* 2018, *5*, doi:10.1109/JIOT.2018.2828144.
[19] Naik, K., Patel, S. An Open Source Smart Home Management System Based on IOT. *Wireless Networks* 2018, doi:10.1007/s11276-018-1884-z.
[20] Park, C., Kim, Y., Jeong, M. Influencing Factors on Risk Perception of IoT-Based Home Energy Management Services. *Telematics and Informatics* 2018, *35*, doi:10.1016/j.tele.2018.10.005.
[21] Rashid, R.A., Chin, L., Sarijari, M.A., Sudirman, R., Ide, T. Machine Learning for Smart Energy Monitoring of Home Appliances Using IoT. In Proceedings of the International Conference on Ubiquitous and Future Networks, ICUFN, 2019, Vol. 2019-July.
[22] T., K., S., C.R., J., J.D.N., K., C. Design of IoT Based Smart Compact Energy Meter for Monitoring and Controlling the Usage of Energy and Power Quality Issues with Demand Side Management for a Commercial

Building. *Sustainable Energy, Grids and Networks* 2021, *26*, 100454, doi:10.1016/j.segan.2021.100454.

[23] H.S., S., Ramaiah, N.S. A Two-Stage Deep Convolutional Model for Demand Response Energy Management System in IoT-Enabled Smart Grid. *Sustainable Energy, Grids and Networks* 2022, *30*, 100630, doi:10.1016/j.segan.2022.100630.

[24] Ahmed, M.A., Chavez, S.A., Eltamaly, A.M., Garces, H.O., Rojas, A.J., Kim, Y.-C. Toward an Intelligent Campus: IoT Platform for Remote Monitoring and Control of Smart Buildings. *Sensors* 2022, *22*, 9045, doi:10.3390/s22239045.

[25] Le, D.N., le Tuan, L., Dang Tuan, M.N. Smart-Building Management System: An Internet-of-Things (IoT) Application Business Model in Vietnam. *Technol Forecast Soc Change* 2019, *141*, doi:10.1016/j.techfore.2019.01.002.

[26] Terroso-Saenz, F., González-Vidal, A., Ramallo-González, A.P., Skarmeta, A.F. An Open IoT Platform for the Management and Analysis of Energy Data. *Future Generation Computer Systems* 2019, *92*, doi:10.1016/j.future.2017.08.046.

[27] Pawar, P., TarunKumar, M., Vittal K., P. An IoT Based Intelligent Smart Energy Management System with Accurate Forecasting and Load Strategy for Renewable Generation. *Measurement (Lond)* 2020, *152*, doi:10.1016/j.measurement.2019.107187.

[28] Tom, R.J., Sankaranarayanan, S., Rodrigues, J.J.P.C. Smart Energy Management and Demand Reduction by Consumers and Utilities in an IoT-Fog-Based Power Distribution System. *IEEE Internet Things J* 2019, *6*, doi:10.1109/JIOT.2019.2894326.

[29] Elsisi, M., Tran, M.Q., Mahmoud, K., Lehtonen, M., Darwish, M.M.F. Deep Learning-Based Industry 4.0 and Internet of Things towards Effective Energy Management for Smart Buildings. *Sensors (Switzerland)* 2021, *21*, doi:10.3390/s21041038.

[30] Rudenko, R., Pires, I.M., Oliveira, P., Barroso, J., Reis, A. A Brief Review on Internet of Things, Industry 4.0 and Cybersecurity. *Electronics (Switzerland)* 2022, *11*.

[31] Kiran, M.P.R.S., Subrahmanyam, V., Rajalakshmi, P. Novel Power Management Scheme and Effects of Constrained On-Node Storage on Performance of MAC Layer for Industrial IoT Networks. *IEEE Trans Industr Inform* 2018, *14*, doi:10.1109/TII.2017.2766783.

[32] Zhang, F., Hu, X., Langari, R., Cao, D. Energy Management Strategies of Connected HEVs and PHEVs: Recent Progress and

Outlook. *Prog Energy Combust Sci* 2019, *73*, 235–256, doi:10.1016/J.PECS.2019.04.002.

[33] Renugadevi, N., Saravanan, S., Naga Sudha, C.M. IoT Based Smart Energy Grid for Sustainable Cites. *Mater Today Proc* 2021, doi:10.1016/J.MATPR.2021.02.270.

[34] Kim, H., Choi, H., Kang, H., An, J., Yeom, S., Hong, T. A Systematic Review of the Smart Energy Conservation System: From Smart Homes to Sustainable Smart Cities. *Renewable and Sustainable Energy Reviews* 2021, *140*, 110755, doi:10.1016/J.RSER.2021.110755.

[35] Abdel-Basset, M., Hawash, H., Chakrabortty, R.K., Ryan, M. Energy-Net: A Deep Learning Approach for Smart Energy Management in IoT-Based Smart Cities. *IEEE Internet Things J* 2021, *8*, 12422–12435, doi:10.1109/JIOT.2021.3063677.

[36] Ejaz, W., Naeem, M., Shahid, A., Anpalagan, A., Jo, M. Efficient Energy Management for the Internet of Things in Smart Cities. *IEEE Communications Magazine* 2017, *55*, 84–91, doi:10.1109/MCOM.2017.1600218CM.

[37] Khajenasiri, I., Estebsari, A., Verhelst, M., Gielen, G. A Review on Internet of Things Solutions for Intelligent Energy Control in Buildings for Smart City Applications. In Proceedings of the Energy Procedia, 2017, Vol. 111.

[38] Kamienski, C.A., Borelli, F.F., Biondi, G.O., Pinheiro, I., Zyrianoff, I.D., Jentsch, M. Context Design and Tracking for IoT-Based Energy Management in Smart Cities. *IEEE Internet Things J* 2018, *5*, doi:10.1109/JIOT.2017.2748037.

[39] Liu, Y., Yang, C., Jiang, L., Xie, S., Zhang, Y. Intelligent Edge Computing for IoT-Based Energy Management in Smart Cities. *IEEE Netw* 2019, *33*, doi:10.1109/MNET.2019.1800254.

[40] Bhatti, G., Mohan, H., Raja Singh, R. Towards the Future of Ssmart Electric Vehicles: Digital Twin Technology. *Renewable and Sustainable Energy Reviews* 2021, *141*.

[41] Agostinelli, S., Cumo, F., Guidi, G., Tomazzoli, C. Cyber-Physical Systems Improving Building Energy Management: Digital Twin and Artificial Intelligence. *Energies (Basel)* 2021, *14*, 2338, doi:10.3390/en14082338.

[42] Opoku, D.G.J., Perera, S., Osei-Kyei, R., Rashidi, M. Digital Twin Application in the Construction Industry: A Literature Review. *Journal of Building Engineering* 2021, *40*, 102726, doi:10.1016/J.JOBE.2021.102726.

[43] Ma, S., Ding, W., Liu, Y., Ren, S., Yang, H. Digital Twin and Big Data-Driven Sustainable Smart Manufacturing Based on Information Management Systems for Energy-Intensive Industries. *Appl Energy* 2022, *326*, 119986, doi:10.1016/J.APENERGY.2022.119986.

[44] Huang, W., Zhang, Y., Zeng, W. Development and Application of Digital Twin Technology for Integrated Regional Energy Systems in Smart Cities. *Sustainable Computing: Informatics and Systems* 2022, *36*, 100781, doi:10.1016/J.SUSCOM.2022.100781.

[45] O'Dwyer, E., Pan, I., Charlesworth, R., Butler, S., Shah, N. Integration of an Energy Management Tool and Digital Twin for Coordination and Control of Multi-Vector Smart Energy Systems. *Sustain Cities Soc* 2020, *62*, 102412, doi:10.1016/J.SCS.2020.102412.

[46] Wang, Y., Kang, X., Chen, Z. A Survey of Digital Twin Techniques in Smart Manufacturing and Management of Energy Applications. *Green Energy and Intelligent Transportation* 2022, *1*, 100014, doi:10.1016/J. GEITS.2022.100014.

[47] Li, X., Liu, H., Wang, W., Zheng, Y., Lv, H., Lv, Z. Big Data Analysis of the Internet of Things in the Digital Twins of Smart City Based on Deep Learning. *Future Generation Computer Systems* 2022, *128*, 167–177, doi:10.1016/J.FUTURE.2021.10.006.

[48] Attaran, M., Celik, B.G. Digital Twin: Benefits, Use Cases, Challenges, and Opportunities. *Decision Analytics Journal* 2023, *6*, 100165, doi:10.1016/j.dajour.2023.100165.

[49] Botín-Sanabria, D.M., Mihaita, S., Peimbert-García, R.E., Ramírez-Moreno, M.A., Ramírez-Mendoza, R.A., Lozoya-Santos, J. de J. Digital Twin Technology Challenges and Applications: A Comprehensive Review. *Remote Sens (Basel)* 2022, *14*.

6

Renewable Energy Integration Amid the 4th Industrial Revolution

Ahmad Eid[1,2] and H. El-Kishky[3]

[1]Department of Electrical Engineering, Faculty of Engineering, Aswan University, Egypt
[2]Department of Electrical Engineering, College of Engineering, Qassim University, Unaizah, Saudi Arabia
[3]Electrical Engineering Department, The University of Texas at Tyler, USA

Emails: ahmadeid@aswu.edu.eg; helkishky@uttyler.edu

Abstract

The 4th Industrial Revolution (4th IR or Industry 4.0) concept is gaining increasing attention around the globe. In this concept, technologies connected to artificial intelligence (AI) applications, the Internet of Things (IoT), big data (BD), and cyber–physical systems (CPSs) are coupled to enhance operational efficiency, productivity, and automation. While emphasizing these benefits, intelligent systems must also be energy-efficient and renewable. This chapter outlines a comprehensive analysis and information about the effects of the 4th IR key applications, especially the impacts of AI, IoT, and BD on renewable energy operation and control. In the second phase of the study, the system loss and voltage regulation are minimized with uncertain demand distribution systems during the 24-hour period. The recently announced multi-objective bonobo optimizer (MOBO) method solves this multi-objective optimization (MOO) problem. The MOO is conducted concurrently with the 24-hour operation of an unpredictable demand distribution system. The methodology is used for a conventional 69-bus distribution network as a baseline. In the meantime, in order to select the best compromise solution (BCS) for the decision-maker (DM) from the Pareto optimal front (POF), two MOO techniques are applied and compared. These techniques include fuzzy-based decision-making (FDM) and the technique for order of preference by similarity to ideal solution (TOPSIS) methods. TOPSIS is a multi-criteria decision analysis method. It is based on the concept that the chosen alternative should have the shortest geometric distance from the positive ideal solution and the longest geometric distance from the negative ideal solution. Statistical analyses are recorded for each technique, highlighting the best one. Moreover, the study shows the effectiveness and the excellent behavior of the MOBO algorithm in solving the multi-objective problems of integrating renewable energy sources into distribution systems.

6.1 Introduction and Background of the 4th IR

6.1.1 Impact of industrial revolutions

The industrial revolution (IR) had a significant impact on the growth of contemporary society because it dramatically altered the society, the economy, and the style of life [1]. Historically, there have been four industrial revolutions (see Figure 6.1). A brief description of each revolution is outlined. The First Industrial Revolution (1st IR) was characterized by the move from manual manufacturing methods to steam- or water-powered equipment. Electricity facilitated the transformation of factories into modern production lines during the Second Industrial Revolution (2nd IR), resulting in high productivity and substantial economic expansion. During the Third Industrial Revolution (3rd IR), field-level computers such as the programmable logic controller and communication technologies were integrated into the manufacturing process, resulting in automated production [2]–[4]. In the 4th IR, cyber–physical production systems can make intelligent decisions through real-time communication and cooperation between manufacturing objects, allowing the mass production of flexible, high-quality, and individualized goods [5].

The 4th IR supports digital technology, also known as the digital revolution. It entails a convergence of technologies that brings the physical and digital worlds closer together. The 4th IR combines cyber–physical systems (CPSs), the Internet of Systems (IoS), the Internet of Things (IoT), and artificial intelligence (AI). Some latest technologies exist now, from autonomous vehicles and drones to software and digital assistants. The 4th IR is anticipated to significantly impact global businesses, economies, and lifestyles [6]. It is also expected to establish new business models based on the vast quantities of data generated throughout the product life cycle. Digital technologies could be utilized to optimize and improve the energy efficiency of production

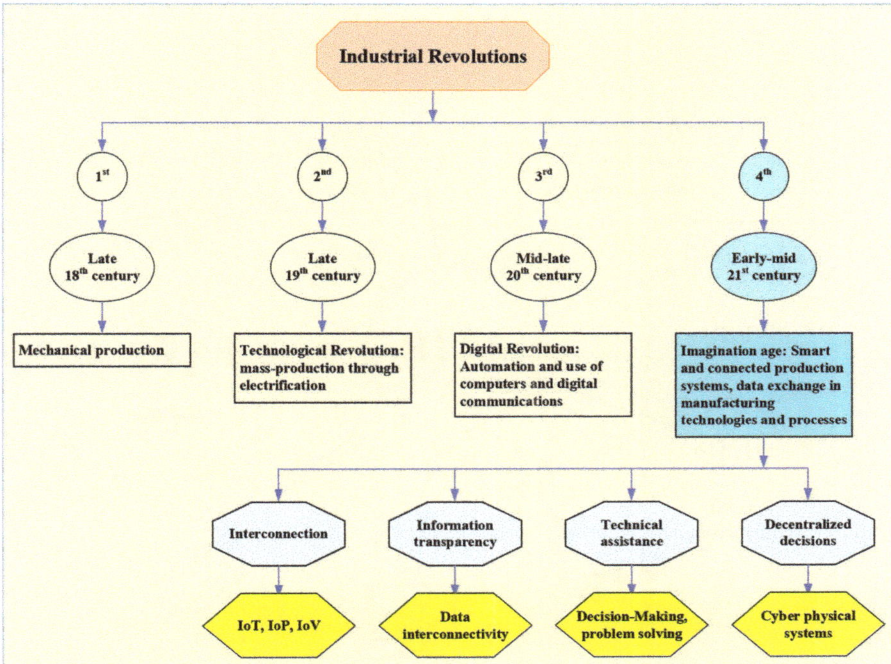

Figure 6.1 Description of industrial revolutions and critical features of the 4th IR.

operations. The 4th IR depends mainly on four design principles, as shown in Figure 6.1. These principles include an interconnection with wireless communications, information transparency, technical assistance to aid humans in decision-making, and decentralized decisions [7], [8].

The primary objective of the 4th IR is to enhance earnings and improve the level of living [9]–[11]. Thus, categories of items and services that can facilitate and accommodate their personal and professional life are established using advanced technologies. Moreover, governments must build effective methods for monitoring and coordinating digital infrastructure because new technology can quickly harm the privacy of individuals. The 4th IR prioritizes environmental protection by encouraging the use of renewable energy sources [12]–[14]. The energy sector, which includes clean and renewable energy, is one of the primary focuses of the 4th IR.

Generally speaking, the 4th IR (Industry 4.0) refers to automation and data exchange methods [6]. These methods are not limited to IoT, CPS, IoS, AI, big data with analytics, additive manufacturing, simulation, augmented reality, system integration, cloud computing, and autonomous robots.

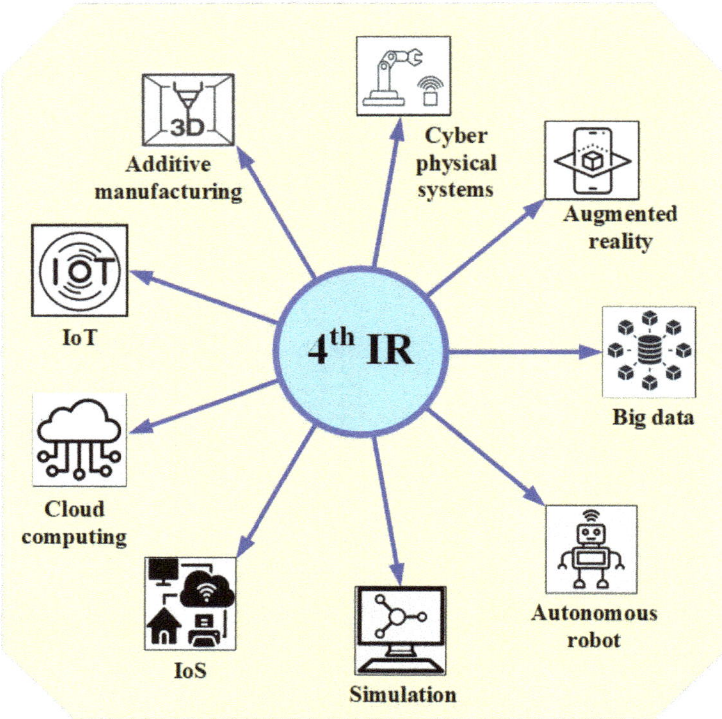

Figure 6.2 Main components of the 4th IR.

A visualization of the main characteristics of the 4th IR is shown in Figure 6.2. It helps to integrate and mix intelligent machines, human actors, physical things, manufacturing lines, and processes across organizational stages to construct new types of technical data and systematic and agile value chains. The 4th IR involves enhanced analytics, connectivity, and automation, allowing organizations to transform everything from factory productivity to product customization with increased market operating efficiency, speed, and new business models. The 4th IR would propel economic expansion [15], [16].

The 4th IR has set the way for the precise deployment of smart power grids to handle the ever-increasing energy demand by incorporating renewable energy resources. Intelligent information processing and advanced information and communication technologies enable energy utilities to monitor and control power generation, transmission, and distribution more efficiently, securely, and cost-effectively in the 4th IR framework [4].

The 4th IR, the high cost of electricity, the depletion of fossil resources, and the problem of global warming have compelled energy utilities and governments to take concrete actions to improve the efficiency and dependability of their existing power grids for delivering high-quality electricity [17]. In addition, it seeks alternative renewable energy sources and their reliable integration into existing power networks in order to satisfy the constantly changing energy needs of the 21st century.

Diverse renewable energy sources are now widely dispersed across the globe and connected to various distribution systems. Unless these enormous dispersed generation systems are monitored and run effectively and securely, the entire power grid will be highly vulnerable [18]. Consequently, in the framework of the 4th IR, effective and secure communications are required for the effective operation of the power grid. Communication, control, and technologically advanced sensors can considerably facilitate the integration of such distributed energy sources into the power grid. The distributed generations are widely used in the smart grid to generate power near or at the load centers [19]–[21]. These sub-systems consist of wind energy, solar power, biomass, and kinetic energy. Renewable energy systems provide clean environmental effects [22]–[24] and decentralized nature for smart grid operation. Moreover, renewable energy sources are gaining popularity over the conventional power grid because of their minimal carbon emissions, improved sustainability, and low cost. In addition, the 4th IR enables the energy business to reorganize its operating environment and enjoy more advanced, intelligent, and complex energy production and distribution equipment. Alternatively, the 4th IR supports the digitalization of the energy demand sector and manufacturing industry. Consequently, the 4th IR offers potential for energy sustainability [22], [23].

6.1.2 Impact of artificial intelligence

Artificial intelligence (AI) facilitates the integration of energy supply, demand, and renewable sources into the power grid and will be autonomously controlled by intelligent software that optimizes decisions and operations [25]. AI will be crucial to reaching this objective. AI is widely employed in solar power generation, hydrogen generation, and demand management control. AI techniques beat conventional models in controllability, big data (BD) management, smart grid, IoT, and computing efficiency [26]–[28]. BD, the construction of a machine learning model, and AI will play a significant role in the future energy market.

Currently, the majority of power networks are ill-equipped to manage the integration of renewable energy sources (RESs). Changes in the properties of RESs (solar power, wind energy, geothermal energy, and hydrogen) make it difficult for utilities to meet their changing load requirements. New advances in AI technologies, including IoT, big data analytics, machine learning, and deep learning, are transforming the energy industry. AI provides effective inverter control of photovoltaic systems and increases the capacity to track power points [29]–[31]. In wind energy, AI is used to track the most significant wind output and adapt to wind and demand variable nature. Optimization algorithms are adopted to integrate RESs and distributed generations to distribution systems to achieve different objectives such as power loss reduction, voltage stability, enhancing voltage profile, and others.

6.1.3 Impact of IoT and BD

The primary function of Internet of Things (IoT) technology is to connect and embed physical devices, automobiles, home appliances, and other components with the electronics, software, sensors, drives, and communication networks to gather and share data [32]. The IoT uses sensors and networking equipment to detect and transmit real-time data for efficient computing and decision-making. Comprehensive data analysis and advanced algorithms can assist customers and inventors in tracking energy usage trends and managing consumption more effectively across different time frames [33]–[38]. Integrating renewable energy and optimizing energy consumption are crucial enablers of sustainable energy transitions and climate change mitigation. Modern technologies of IoT and BD have numerous uses in the energy sector, specifically in energy production, transmission and distribution, and consumption. The IoT can improve energy efficiency, increase the proportion of renewable energy, and reduce the environmental implications of energy consumption.

The IoT and BD have various applications in distribution systems in the context of the 4th IR. Critical applications of IoT and BD are listed in Figure 6.3, covering different areas, including smart grids, microgrids, renewable energy integration, and demand response. Moreover, metering infrastructure, battery energy storage systems management and control, advanced metering infrastructure, and smart building buildings are other examples of IoT and BD applications for distribution systems. The primary roles of IoT and BD are to collect data, analyze them, and hence control the operation of different networks or industries.

6.2 Mathematical Formulations of the Problem

Distributed generations (DGs) are fast gaining popularity due to their efficacy in reducing system capital costs and resolving capacity issues. They

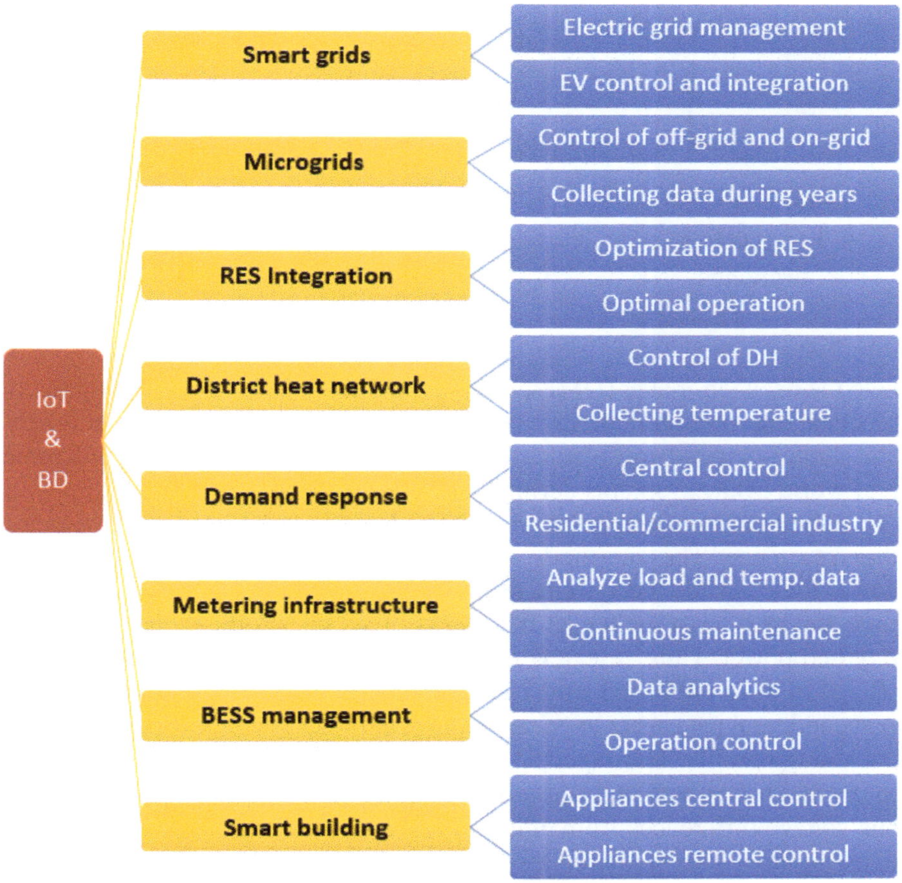

Figure 6.3 Key applications of IoT and BD in distribution systems.

also require little space, are quick to build, and cut greenhouse gas emissions. Increased demand, liberalization of the electricity market, and worries about greenhouse emissions have contributed to a recent surge in interest in DGs. DG sizing and allocation are critical to system efficiency, dependability, and power quality. There are two ways to tackle multi-objective problems. First, using weighting factors, the multi-objective problem becomes a single-objective problem with a single dependent solution. Even a tiny change in the parameter's weight affects the DG's size and placement forecast. The second strategy is to create non-dominated Pareto optimal solutions using meta-heuristic algorithms. The primary purpose of the multi-objective optimization (MOO) method is to discover diverse Pareto optimal solutions to

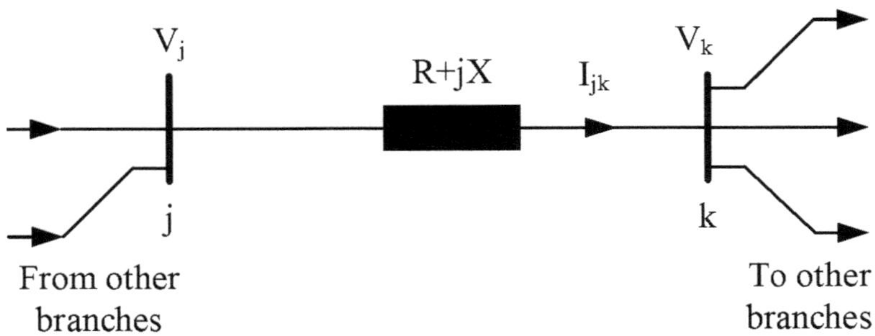

Figure 6.4 A connected branch between two buses *j* and *k*.

help the decision-maker (DM) select the most acceptable option among many practical and optimum alternatives. Numerous multi-objective evolutionary and nature-inspired meta-heuristic algorithms have been created to handle real-world MOO problems during the last three decades.

The branches of distribution systems have a more significant resistance than reactance. The enormous amount of branch resistance leads to high power losses inside the system, resulting in a system deficiency. At the same time, the loading of the feeder increases to feed the system demand with the required power. Thus, researchers tried to decrease the system power losses (P_{LOSS}) in most published works in the first step of design and optimization. A second vital parameter to enhance the quality of the distribution systems is the voltage profile. The expected voltages at all buses should approach unity to operate the system and equipment attached to it properly. The sum of all bus voltage differences from unity is called the total voltage deviation (V_{TD}) of the system. This study reduces these vital factors (objective functions) by using the multi-objective bonobo optimizer (MOBO) algorithm. The definition of these objectives is as follows.

6.2.1 Single-objective optimization

For any branch, as shown in Figure 6.4, connecting two buses *i*, *j* with an impedance of $R + jX$, the power loss per phase is calculated as follows:

$$P_{LOSS}^{Bch} = \left|I_{jk}\right|^2 \times R \tag{6.1}$$

where P_{LOSS}^{Bch} represents the branch power loss; I_{jk} is the branch current.

The total power loss of the system is the summation of power losses of all branches inside the system as follows:

$$P_{\text{LOSS}} = \sum_{i=1}^{N_{\text{Bch}}} P_{\text{LOSS}}^{\text{Bch},i} \tag{6.2}$$

where P_{LOSS} is the total system active losses; N_{Bch} is the number of branches of the system.

The first fitness (objective) function is expressed as follows:

$$f_1 = \min(P_{\text{LOSS}}). \tag{6.3}$$

The voltage deviation of the voltage at bus j is calculated as follows:

$$V_D = \left| \left| V_j \right| - 1.0 \right| \tag{6.4}$$

where V_D is the voltage difference between the voltage at any bus j, V_j and unity.

The total voltage deviation (V_{TD}) of all bus voltages from unity is

$$V_{\text{TD}} = \sum_{i=1}^{N_B} V_{D,i} \tag{6.5}$$

where N_B is the number of buses in the system.

The second fitness function is expressed as follows:

$$f_2 = \min(V_{\text{TD}}). \tag{6.6}$$

6.2.2 Multi-objective optimization (MOO)

Many engineering design issues are multi-objective in nature, as they frequently entail optimizing multiple design objectives. These design objectives may collide with a specific system design's technical and economic performance criteria. Creating a MOO problem is necessary to investigate the tradeoffs between these competing design objectives and explore design alternatives. In order to assist the investigation of various multi-objective formulations, we can add Pareto optimality notions into optimization algorithms that require designer input.

MOO issues confront numerous engineering disciplines. Even if the concept behind these challenges is basic and well-established, implementing a solution is not an easy undertaking. The widespread usage of evolutionary algorithms to discover potential solutions provides a discrete representation of non-dominant solutions, a Pareto set. Although it is pretty interesting to be aware of the non-dominant alternatives, a further criterion is required to choose a solution for deployment.

The fitness (cost) has more than one objective function in a multi-objective problem. A MOO can be expressed as the following minimization problem [39]:

$$\text{Minimize}: F(x) = \{f_1(x), f_2(x), \ldots, f_o(x)\} \tag{6.7}$$

$$\text{Subject to}: \; g_i(x) \geq 0, \; i = 1, 2, \ldots, m \tag{6.8}$$

$$h_i(x) = 0, \; i = 1, 2, \ldots, p \tag{6.9}$$

$$\text{LB}_i \leq x_i \leq \text{UB}_i, \; i = 1, 2, \ldots, n \tag{6.10}$$

$$x = \{x_1, x_2, x_3, \ldots, x_{n-1}, x_n\} \tag{6.11}$$

where f_i is the ith objective function; o is the number of objectives; x is a vector that contains all decision variables of the problem; m, p represent the number of inequality and equality constraints; n represents the number of variables; LB_i, UB_i are the lower and upper bounds of the ith variable.

The outcome of a single- or multi-objective optimum design process is a solution that best meets the formally and/or informally established objectives. If numerous optimization criteria are specified inside the optimization issue, a compromise solution must be determined from a group of multiple optimal solutions. Pareto optimality is a measure of efficiency in multi-objective contexts where multiple competing objectives must be considered throughout the optimization process. A design is Pareto optimum if no other design can increase the value of any of its objective criteria without degrading at least one other criterion.

The mathematical formulation of dominance between two solutions when x_1 dominates x_2 if [40]:

$$f_i(x_1) \leq f_i(x_2) \; \forall \, i \in \{1, \ldots, m\} \text{ and } \exists j \in \{1, \ldots, m\} \, | \, f_i(x_1) < f_i(x_2). \tag{6.12}$$

Consequently, a design x^* is deemed non-dominated or Pareto optimum if no other viable design is superior to it. The Pareto front is the optimal tradeoff between all objectives and is composed of all non-dominated designs.

In this chapter, a bi-objective optimization problem is solved using the MOBO algorithm. The first objective (fitness) is the reduction of the system losses, and the second objective is the reduction of the total voltage deviation. Thus, the bi-objective function is

$$F_{\text{obj}} = \{f_1, f_2\}. \tag{6.13}$$

6.2.3 Problem and system constraints

The associated constraints of the problem include the acceptable voltage limits at every bus of the system, the amount of installed DG units and their locations, and the consistency of the power flow equations at every step of optimization. These constraints are briefly explained as follows.

- Voltage limits

The voltage at any bus of the distribution system should lie between acceptable limits as follows:

$$V_{\min} \leq V_i \leq V_{\max} \tag{6.14}$$

where $V_{\min}$, $V_{\max}$ are the lower and upper acceptable voltages of any bus voltage V_i.

- Installed DG size

The installed DG size (either single or multiple) does not exceed the total system demand, either single or multiple units, as follows:

$$\sum_{i=1}^{N_{\text{DG}}} P_{\text{DG},i} \leq \sum_{k=1}^{N_B} P_{D,i} \tag{6.15}$$

where $P_{\text{DG},i}$ is the power of the DG unit installed at bus i; $P_{D,i}$ is the demand power at the same bus; N_{DG} is the number of DG units.

- Candidate buses for DG units

The MOBO algorithm locates the DG units at their optimal locations, and these locations can be any bus of the system starting from 1 to N_B.

- Power flow equations

At any iteration of the algorithm, the power flow equations are fulfilled. The total power supplied by utility and DG units always equals the system demand and losses as follows:

$$P_{\mathrm{UT}} + \sum_{i=1}^{N_{\mathrm{DG}}} P_{\mathrm{DG},i} = P_{\mathrm{LOSS}} + \sum_{k=1}^{N_B} P_{D,i} \tag{6.16}$$

$$Q_{\mathrm{UT}} + \sum_{i=1}^{N_{\mathrm{DG}}} Q_{\mathrm{DG},i} = Q_{\mathrm{LOSS}} + \sum_{k=1}^{N_B} Q_{D,i} \tag{6.17}$$

where P_{UT}, Q_{UT} are the active and reactive powers supplied by the utility; $P_{\mathrm{DG},i}$, $Q_{\mathrm{DG},i}$ are active and reactive powers of the DG installed at bus *i*; P_{LOSS}, Q_{LOSS} are the system active and reactive losses; $P_{D,i}$, $Q_{D,i}$ are the demand powers at the same bus.

When the DG operates at unity power factor (UPF), it injects just active power into the system, whereas in the case of optimal power factor (OPF), it injects both active and reactive powers.

6.3 Decision-making Techniques

In contrast to single-objective optimization, MOO provides many non-dominated alternatives for constructing a curve. MOO employs many strategies to discover the best compromise solution (BCS) among the set of non-dominated points. Various scholars have attempted to categorize strategies for solving MOO problems based on various criteria [39]–[43]. Depending on how the DM needs preferences, they proposed the following four classes [44]: no-preference techniques, *a priori* methods, *a posteriori* methods, and interactive approaches. While the DM's opinions are considered before the solution process in *a priori* preference articulation methods, they are not considered in no-preference methods. None of the DM's preferences are considered in posterior approaches. After the creation of the Pareto set, the DM chooses a suitable solution [21]. In contrast, the interactive technique allows the DMs to continuously modify throughout the search process.

In particular, this study adopted two techniques for the interactive preference approach in which the BCS is chosen based on the optimization process without the intervention of the DM. These techniques are the FDM approach [20], [45]–[52], and technique for order preference by similarity to ideal solution (TOPSIS). Few research works applied the TOPSIS approach [53]–[57] in power distribution systems. Both techniques are adopted in this study and elaborated as follows.

The objective functions must be normalized in all systems to produce roughly identical weighting effects. The normalization in this study is based on the base scenario of system losses or voltage variations without integrating any DG units.

$$f_1^{\mathrm{nor}} = \frac{f_1^{\mathrm{DG}}}{f_1^{\mathrm{Ba}}} \tag{6.18}$$

$$f_2^{\mathrm{nor}} = \frac{f_2^{\mathrm{DG}}}{f_2^{\mathrm{Ba}}} \tag{6.19}$$

where f_1^{nor}, f_2^{nor} are the first and second normalized objective functions; f_1^{DG}, f_2^{DG} are the corresponding values with installing DG units; f_1^{Ba}, f_2^{Ba} are the base case values of the two objective functions.

6.3.1 FDM approach

The MOBO algorithm determines a complete set of candidate solutions called Pareto optimal front (POF). The BCS point without the intervention of the DM is determined in this study using FDM and TOPSIS approaches. The simple algorithm of the FDM approach is summarized in Figure 6.5 and the following two equations [20], [45]–[52].

$$A_i^k = \begin{cases} 1, & f_i < f_i^{\min} \\ \dfrac{f_i^{\max} - f_i}{f_i^{\max} - f_i^{\min}}, & f_i^{\min} \le f_i \le f_i^{\max} \\ 0, & f_i > f_i^{\max} \end{cases} \tag{6.20}$$

$$A^k = \frac{\sum_{i=1}^{N_{\mathrm{ob}}} A_i^k}{\sum_{k=1}^{N_{\mathrm{ns}}} \sum_{i=1}^{N_{\mathrm{ob}}} A_i^k} \tag{6.21}$$

where f_i, $f_i^{\min}$, $f_i^{\max}$ are the objective function i with a non-dominated solution k and its minimum and maximum objective values, respectively; N_{ob} is the number of the considered objectives; N_{ns} is the total number of the non-dominated solution points; A^k vector is the normalized values of A_i^k for each solution k.

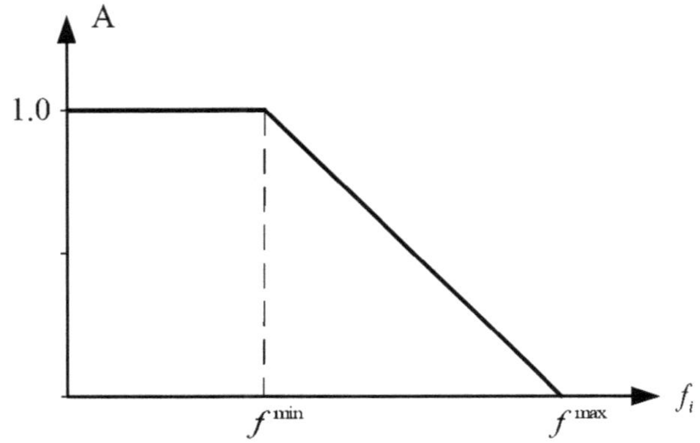

Figure 6.5 Membership of objective functions in the FDM approach.

The vector A^k is arranged in descending order, and the first value (the biggest) is considered the BCS of the problem amongst the non-dominated solution of POF.

6.3.2 TOPSIS approach

The fundamental procedure of TOPSIS consists of normalizing the decision matrix, selecting the positive and negative ideal solutions, computing the distances between each solution and the ideal solutions, and determining the optimal solution based on the degree of proximity to the positive ideal solution. The TOPSIS method is readily applicable to multi-objective decision-making situations because there are no restrictions on data distribution or sample size. In the proposed model, the number of Pareto solutions are determined by MOBO before TOPSIS determines the best solution. The following steps summarize the TOPSIS approach [53]–[57].

1. If the problem has n Pareto solutions (evaluation alternatives) and m evaluation objectives, then the decision matrix is built as an $n \times m$ matrix:

$$S = \begin{bmatrix} x_{11} & x_{12} & \cdots & x_{1m} \\ x_{21} & x_{22} & \cdots & x_{2m} \\ \vdots & \vdots & \ddots & \vdots \\ x_{n1} & x_{n2} & \cdots & x_{nm} \end{bmatrix}. \tag{6.22}$$

2. Normalizing the decision matrix into a non-dimensional matrix by normalizing each element:

$$N = \begin{bmatrix} v_{11} & v_{12} & \cdots & v_{1m} \\ v_{21} & v_{22} & \cdots & v_{2m} \\ \vdots & \vdots & \ddots & \vdots \\ v_{n1} & v_{n2} & \cdots & v_{nm} \end{bmatrix}. \tag{6.23}$$

Each element in the matrix is normalized as follows:

$$v_{ij} = \frac{x_{ij}}{\sqrt{\sum_{i=1}^{n} \left(x_{ij}\right)^2}}. \tag{6.24}$$

3. To obtain the weighted decision matrix Y, express the relative significance of each criterion using a weighting vector ω.

$$\omega = \left[\omega_1, \omega_2, \omega_3, \ldots, \omega_m\right]^T, \sum_{i=1}^{m} \omega_i = 1 \tag{6.25}$$

$$Y_{ij} = \omega_j \times N_{ij} \tag{6.26}$$

where ω_j is the weight of the jth objective.

4. Determine positive (Y^+) and negative (Y^-) ideal solution vectors:

$$Y^+ = \left(y_1^+, y_2^+, \ldots, y_m^+\right) = \left(\max y_{ij} \middle| j \in j^+, \min y_{ij} \middle| j \in j^-\right) \tag{6.27}$$

$$Y^- = \left(y_1^-, y_2^-, \ldots, y_m^-\right) = \left(\min y_{ij} \middle| j \in j^+, \max y_{ij} \middle| j \in j^-\right) \tag{6.28}$$

$$1 \le i \le n \tag{6.29}$$

where j^+ is a set of benefit attributes (the bigger, the better); j^- represents a set of cost attributes (the smaller, the better).

5. Calculate the distance between each alternative and the ideal solution as follows:

$$D_i^+ = \sqrt{\sum_{j=1}^{m} \left(y_{ij} - y_j^+\right)^2}, \quad i = 1, 2, \ldots, n \tag{6.30}$$

$$D_i^- = \sqrt{\sum_{j=1}^{m} \left(y_{ij} - y_j^-\right)^2}, \quad i = 1, 2, \ldots, n. \tag{6.31}$$

6. Calculate the relative closeness θ_i of the solution to the ideal one:

$$\theta_i = \frac{D_i^+}{D_i^+ + D_i^-}, \quad i = 1, 2, \ldots, n \tag{6.32}$$

The optimal solution or BCS is the maximum value of θ_i.

6.4 Multi-objective Bonobo Optimizer

The Bonobo optimizer is inspired by the social behavior and reproduction techniques of bonobos. Bonobos employ a social method known as fission–fusion, which involves the formation of numerous groups of varying size and composition within a community, followed by a brief reunification with their own communities. The Bonobo optimizer, like other heuristics, is an algorithm based on a population. Each individual in the population is referred to as a bonobo. The highest-ranking bonobo in the population dominance hierarchy is known as the alpha-bonobo. If the iteration enhances the alpha-bonobo, the phase is considered favorable. In contrast, if the alpha-bonobo does not progress, this is known as a negative phase. Detailed descriptions and governing equations are found in [45] and [58]. The adopted MOBO is based on the NSGA-II algorithm [59]. It inherits the principles of non-dominated sorting, population rank assignment, and crowding distance from the NSGA-II algorithm.

6.5 Simulations and Discussions

The MOBO algorithm optimizes the 69-bus distribution system to decrease both the system losses and the total voltage deviations. The system losses (P_{LOSS}) are the sum of all system branches every hour of operation when installing single or multiple DG units into the system. The total voltage deviation (V_{TD}) represents the accumulated differences at all system buses from the unity. Different (eight) case studies are simulated and analyzed as follows:

- MOO with integrating a single DG unit (UPF and OPF) with fixed and variable demand.
- MOO with integrating three DG units (UPF and OPF) with fixed and variable demand.

Using two techniques, every case study is optimized to identify the BCS generated from the Pareto optimal set. The 69-bus system is composed of 68 branches and 69 buses with a 12.66-kV rating and a combined load rating

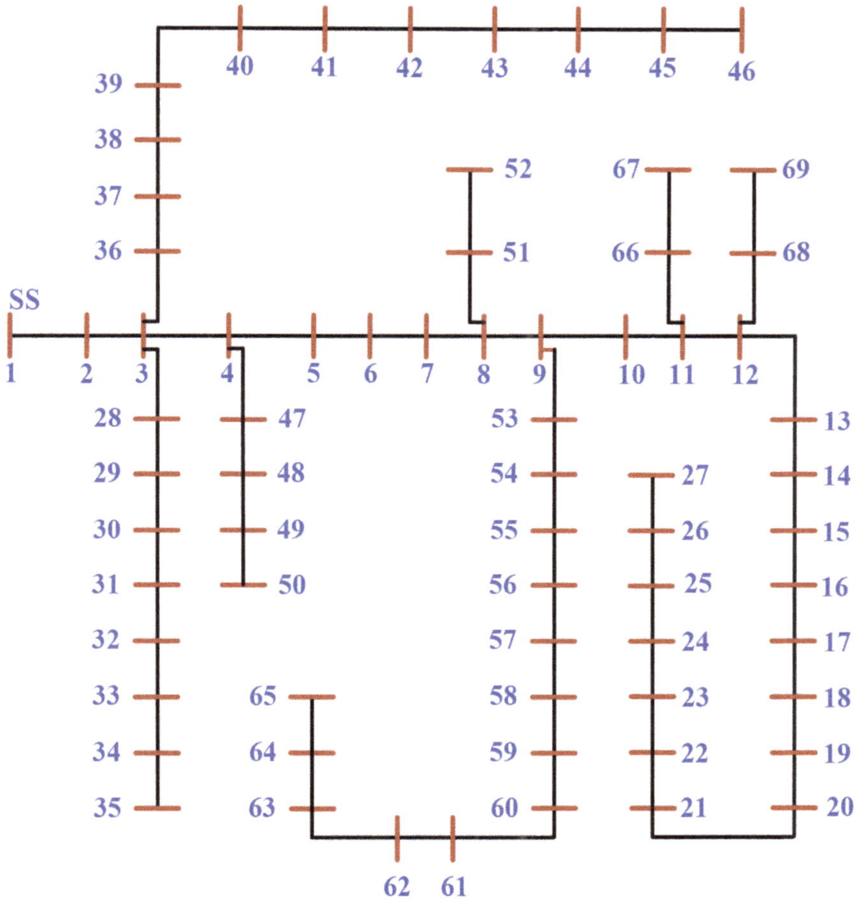

Figure 6.6 Schematic diagram of the 69-bus system.

of 3801.89 kW and 2694.1 kVar [60]. The power loss is 224.95 kW for the base case. The system layout is shown in Figure 6.6, and its complete data are listed in the Appendix. The forecasted load power profile for the next 24 hours [61] is shown in Figure 6.7.

6.5.1 Allocation of a single-DG at full demand

6.5.1.1 With UPF-DG

This section of the study demonstrates the system behavior during the fixed total load demand. At the same time, the MOBO algorithm optimizes the single DG unit at UPF to reduce the system losses and voltage deviations. Both

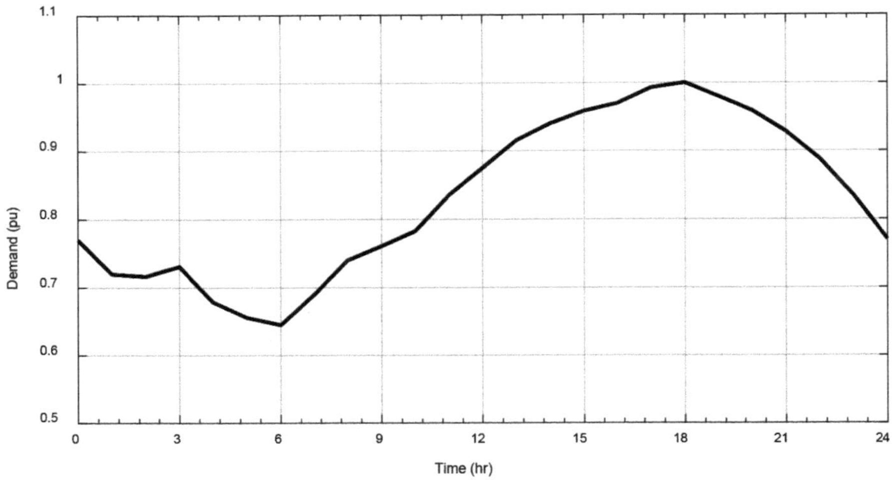

Figure 6.7 Daily load profile.

fuzzy decision-making (FDM) and TOPSIS approaches are used to capture the BCS point of operation.

As the optimization solutions are usually not fixed and probabilistic, many runs are adopted, and the average of solutions is taken for comparison purposes. The MOBO algorithm determines the best operation points in a series of non-dominated points known as the Pareto optimal front (POF).

The POF points represent the best operation points that the distribution system operator (DSO) can choose depending on the system setting and conditions. If the DSO decides to decrease the system losses at the expense of the voltage profile, he/she can choose an operating point at the left of the curve shown in Figure 6.8. On the other hand, if the DSO wishes to enhance the voltage profile (and hence decrease the total voltage deviation), he/she must choose an operating point to the right of the curve. For the automatic and equal opportunity selection of the best operation point, both FDM and TOPSIS can determine it efficiently. The two approaches determine the BCS points that are very close, as shown. The corresponding coordinates of the BCS points from FDM and TOPSIS are (0.42296, 0.32987) and (0.41594, 0.33495), as shown in Figure 6.8. It is worth mentioning that as the population size of bonobos increases, these points get closer or coincide.

6.5.1.2 With OPF-DG

When allocating a DG unit operating with an OPF, the DG injects active and reactive powers into the distribution system at the DG bus. Hence, the power

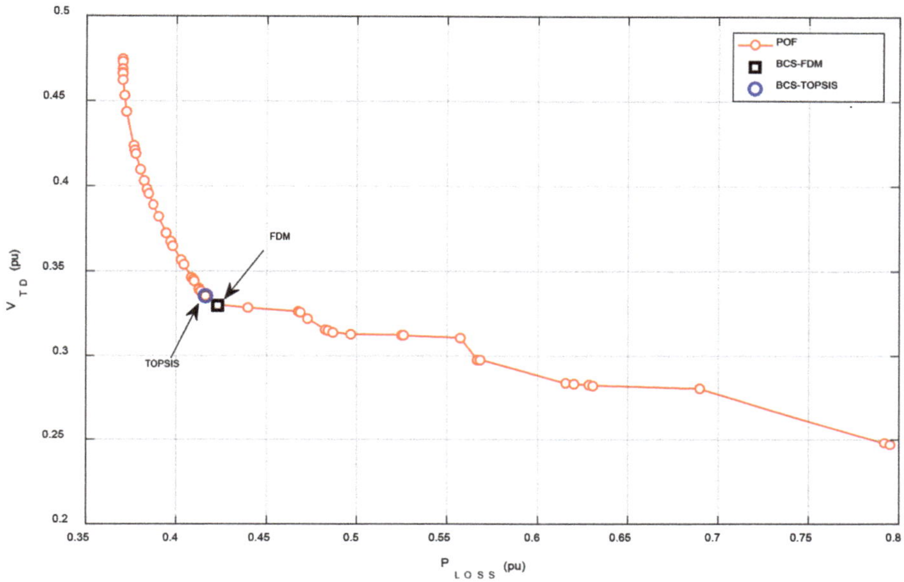

Figure 6.8 The POF and BCS points for a single DG allocation at UPF.

flow in the branches decreases, and, consequently, the losses decrease in the system. Thus, with the OPF DG unit, the losses are expected to decrease, and the voltage profile is enhanced, as shown in Figure 6.9. The BCS points for FDM and TOPSIS approaches coincide at (0.10429, 0.30484). As can be seen, the FDM and TOPSIS coincide, with much lower losses than the UPF case. The optimal size of the DG is 1233.4 kW with an OPF of 0.756.

6.5.2 Three-DG allocation at full demand

6.5.2.1 With UPF-DGs

In this section, three DG units operating at UPF are optimized and allocated into the 69-distribution system. The MOBO algorithm efficiently determines the optimal sizes and sites of the three DG units. The optimal sizes are 935.14, 437.39, and 1985.7 kW at buses 11, 18, and 61. When the DG units increase, the voltage profile enhances due to the injection of the real power at the three buses of the system. The injected power lessens the current in the feeders, and, hence, the voltage drops decrease. Consequently, the bus voltages rise. The Pareto set as well as the BCS of the FDM and TOPSIS approaches are shown in Figure 6.10. Compared to the FDM, which attempts to reduce power loss at the expense of voltage deviations,

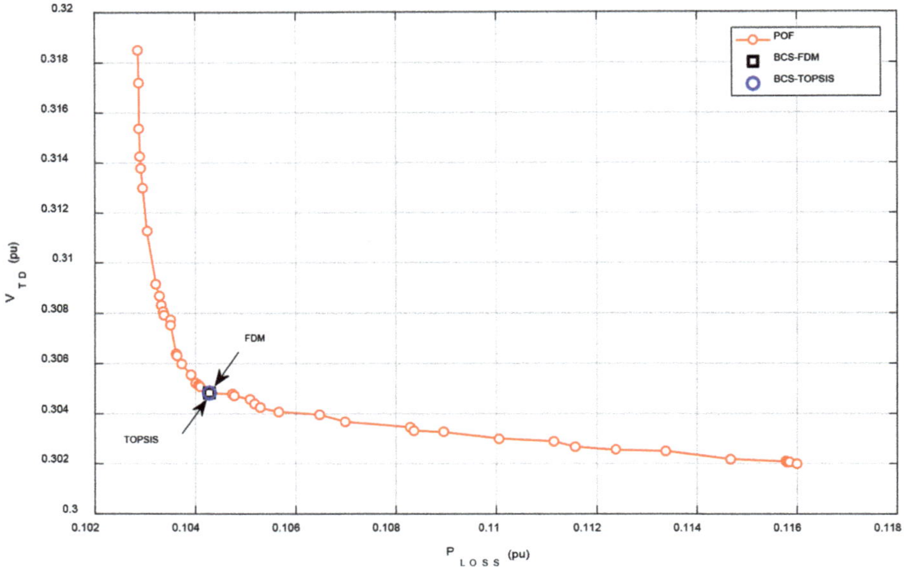

Figure 6.9 The POF and BCS points for a single DG allocation at OPF.

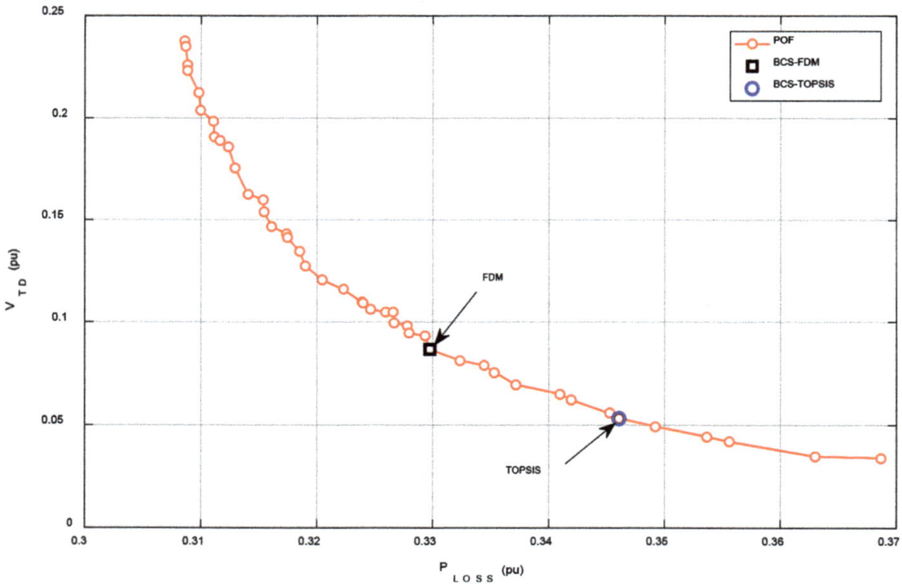

Figure 6.10 The POF and BCS points for the three-DG unit allocation at UPF.

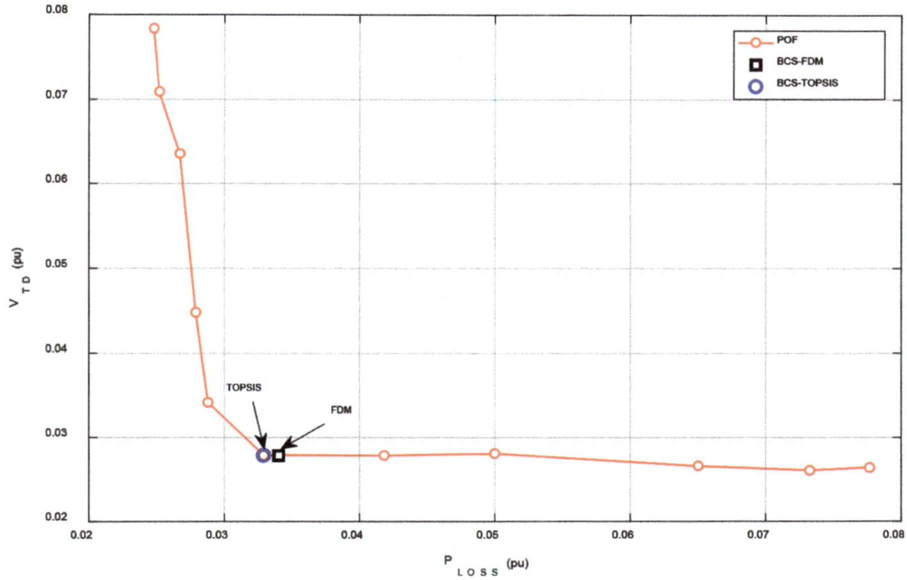

Figure 6.11 The POF and BCS points for the three-DG unit allocation at UPF.

the TOPSIS strives to reduce voltage deviations at system loss. The recorded BCS point coordinates are (0.3298, 0.0866) and (0.3461, 0.0534) for FDM and TOPSIS, respectively.

6.5.2.2 With OPF-DGs

This section optimizes and assigns three DG units operating at OPF to the 69-distribution system. The MOBO algorithm identifies the best sizes and locations of the three DG units. The optimal locations are buses 11, 18, and 61, and the ideal sizes are 641.22, 289.47, and 1745.40 kW, with optimal power factors of 0.8187, 0.8306, and 0.8081, respectively. Due to the injection of real power at the three buses of the system, the voltage profile improves as the DG units rise. The injected power reduces the current in the feeders, hence reducing the voltage dips. Due to this, the bus voltages increase. Figure 6.11 displays the Pareto set in addition to the BCS of the FDM and TOPSIS techniques. The BCS points for both approaches are very close. The BCS point coordinates for FDM and TOPSIS are (0.0340, 0.0278) and (0.0329, 0.0278). The operation of DG with OPF decreases both objectives of loss and voltage deviations.

The voltage profiles for the DG UPF cases compared to the base case (without DG) are investigated. There is an immediate improvement in the

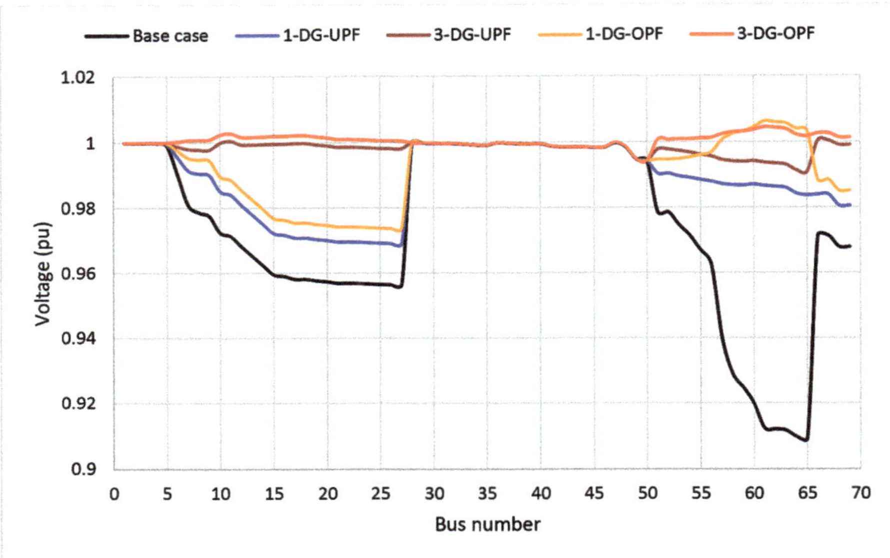

Figure 6.12 Effect of DG allocations on voltage profile.

voltage profile when installing the DG units. Moreover, the voltage has a better profile with increasing the number of DG units. At the same time, allocating the DG units does not violate the acceptable voltage limits (0.95 and 1.05 pu). The minimum voltage has been improved from 0.9692 pu for a single DG to 0.9908 pu for the three DG units operating at UPF compared to the value of 00.9092 pu of the base case. More improvements in voltage profile are achieved when the installed DG operates with OPF, as shown in Figure 6.12. The minimum and maximum voltages for all case studies are listed in Table 6.1. The improvements with OPF are apparent.

6.5.3 Allocation of a single-DG at variable demand

In this section, the MOBO algorithm optimizes a single-DG unit attached at an ideal location to a 69-bus distribution system to minimize two objective functions: system power loss and the total deviations of their bus voltages from unity. Consideration is given to both the UPF and OPF case studies.

6.5.3.1 With UPF-DG

The integration of a single UPF DG with the 69-bus distribution system is located at bus 61, the optimal location with the typical demand, and reduces the total system losses [20]. For the MOO of both power loss and voltage

Table 6.1 Effect of DG allocations on voltage limits.

Limits	Base case	UPF DG		OPF DG	
		One DG	Three DGs	One DG	Three DGs
Min voltage	0.9092	0.9692	0.9908	0.9738	0.9943
Max voltage	1.0000	1.0000	1.0005	1.0064	1.0046

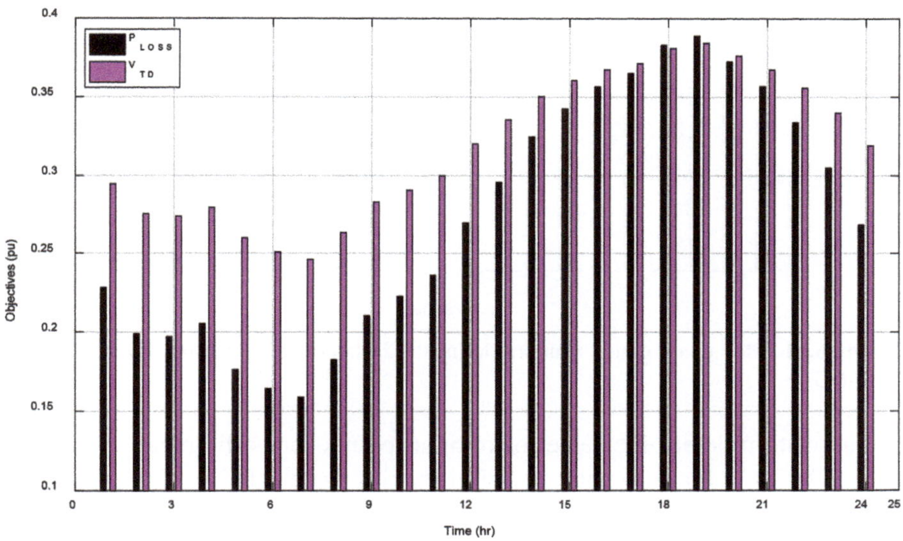

Figure 6.13 Best compromise solutions during the 24-hour operation using FDM for one UPF-DG.

deviation, the DG unit is located at the same bus 61. The MOBO algorithm is adopted to reduce losses and enhance the voltage profile during the day. For every hour, the algorithm determines a set of optimal points for both objectives (P_{LOSS}, V_{-TD}). The BCS for automatic decision-making is selected from this dataset. Both fuzzy decision-making (FDM) and TOPSIS approaches are adopted and listed in Figures 6.13 and 6.14. As can be seen, the BCS depends on the demand and varies from hour to hour. The reductions in power loss are more significant than those in voltage deviations with the FDM approach for most of the day. The opposite situation is with the TOPSIS approach, where more reductions exist with voltage deviation.

6.5.3.2 With OPF-DG

By allocating a single DG with OPF, the MOBO algorithm is implemented to reduce losses and enhance the voltage profile during the day. For every hour,

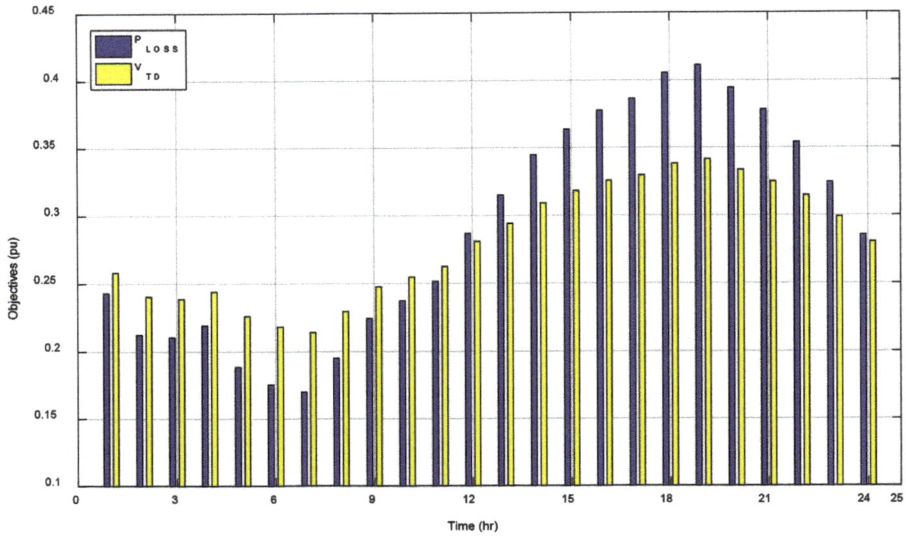

Figure 6.14 Best compromise solutions during the 24-hour operation using TOPSIS for one UPF-DG.

the algorithm determines a set of optimal points for both objectives (P_{LOSS}, V_{TD}). The BCS for automatic decision-making is selected from this dataset. Both FDM and TOPSIS approaches are adopted and listed in Figures 6.15 and 6.16. As can be seen, the BCS depends on the demand and varies from hour to hour. The reductions in power loss are more significant than those in voltage deviations with the FDM approach for all the day. The same situation is with the TOPSIS approach, where more reductions exist with power losses. If we look carefully at both figures, we will find that the power losses and voltage deviations decrease with OPF-DG compared to the previous case of UPF-DG. The general profile for both objectives follows the daily load curve.

The average and standard deviations of both objectives for UPF- and OPF-DG are listed in Table 6.2. In the case of UPF-DG, the average percentage reductions of P_{LOSS} and V_{TD} are 27.26% and 31.86%, respectively, for the FDM approach. On the other hand, these average reductions are 28.98% and 28.02% for the TOPSIS approach. The values of Std are lower for V_{TD} than P_{LOSS} for both approaches. The average and standard deviation in the case of OPF-DG are lower than the UPF-DG case. The values of these parameters are comparable for both approaches of FDM and TOPSIS. The optimal size of the DG unit differs from hour to hour according to the demand.

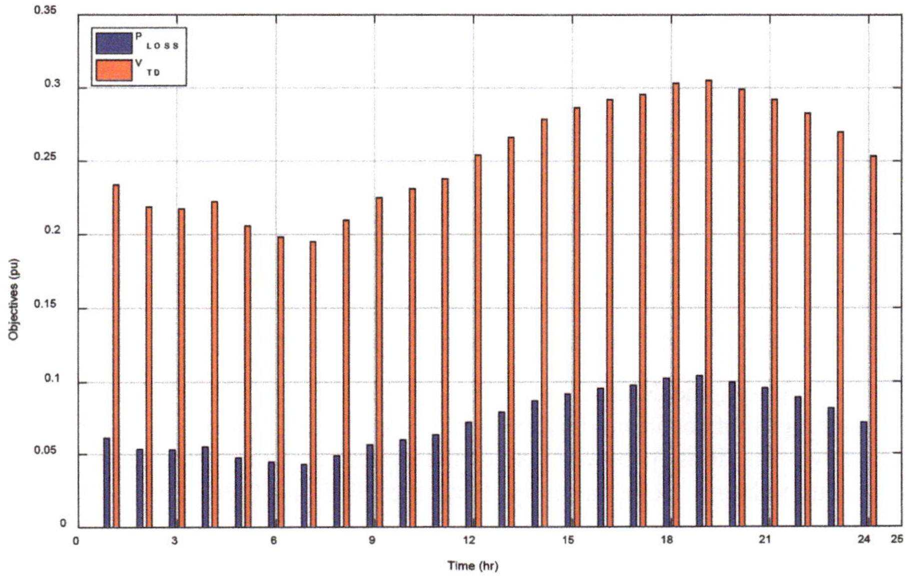

Figure 6.15 Best compromise solutions during the 24-hour operation using FDM for one OPF-DG.

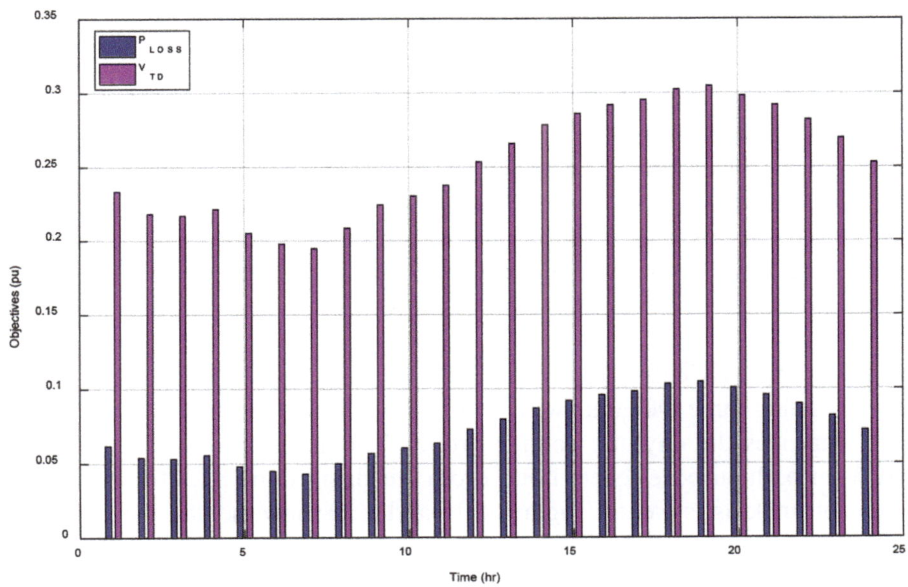

Figure 6.16 Best compromise solutions during the 24-hour operation using TOPSIS for one OPF-DG.

Table 6.2 Average and standard deviation parameters for FDM and TOPSIS methods with One DG.

DG PF	Parameter	FDM		TOPSIS	
		P_{LOSS}	V_{TD}	P_{LOSS}	V_{TD}
UPF	Ave	0.27262	0.31862	0.28982	0.28023
	Std	0.07767	0.04607	0.08158	0.04232
OPF	Ave	0.07309	0.25306	0.07360	0.25255
	Std	0.02070	0.00750	0.00422	0.00751

6.5.4 Allocation of three DGs at variable demand

In this section, the MOBO algorithm optimizes three DG units attached at their optimal locations to the 69-bus distribution system to reduce two objective functions of power loss of the system and the total deviations of their bus voltages from the unity. Both the UPF and OPF case studies are considered.

6.5.4.1 With UPF-DG types

The MOBO algorithm first determines the optimal locations of the three DG units at the base case, and then they are fixed in their position during optimization. The obtained optimal locations are buses 11, 18, and 61. The MOBO algorithm is implemented to reduce losses and improve the voltage profile throughout the day. The program determines a set of optimal positions for both purposes every hour. The BCS for automatic decision-making is identified from this dataset. Both fuzzy decision-making (FDM) and TOPSIS techniques are utilized, as shown in Figures 6.17 and 6.18. As can be observed, the BCS fluctuates from hour to hour based on demand. Throughout most of the day, the reductions in power loss are more significant than those in voltage variances with the FDM method. With the TOPSIS method, voltage variation results in more significant decreases.

6.5.4.2 With OPF-DG types

By assigning three DG units with OPF, the MOBO algorithm reduces losses and improves the daytime voltage profile. The algorithm determines a set BCS for both purposes every hour. From this dataset, the BCS for automatic decision-making is picked. Figures 6.19 and 6.20 show the daily distribution of BCS of FDM and TOPSIS techniques. As can be observed, the BCS fluctuates from hour to hour based on demand. The FDM method reduces power loss significantly more than it reduces voltage fluctuations throughout the day. Similarly, power loss reductions are increased with the TOPSIS strategy.

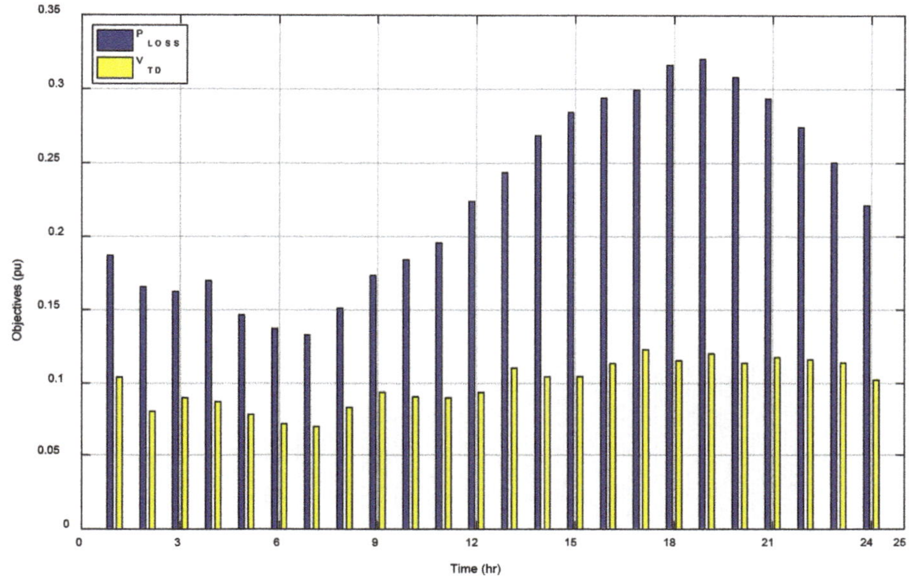

Figure 6.17 Best compromise solutions during the 24-hour operation using FDM for three UPF-DGs.

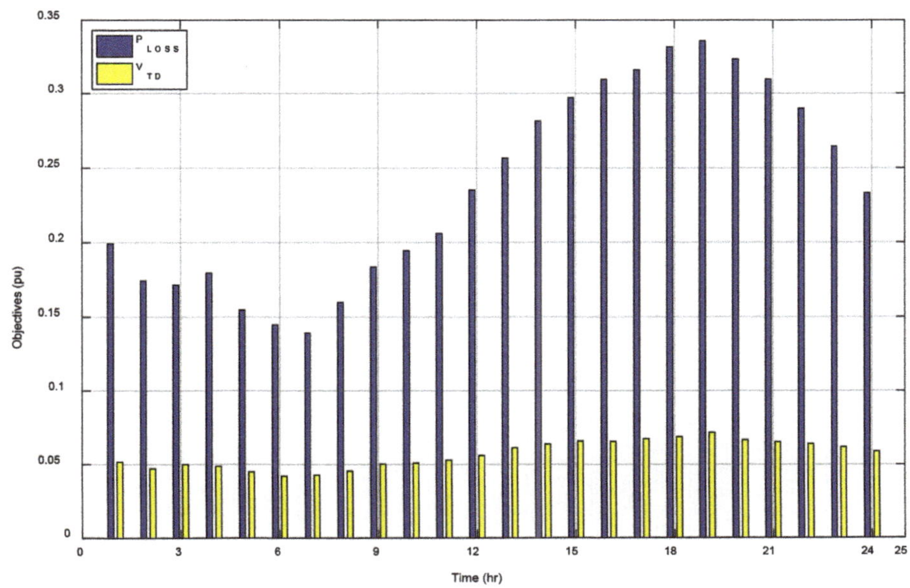

Figure 6.18 Best compromise solutions during the 24-hour operation using TOPSIS for three UPF-DGs.

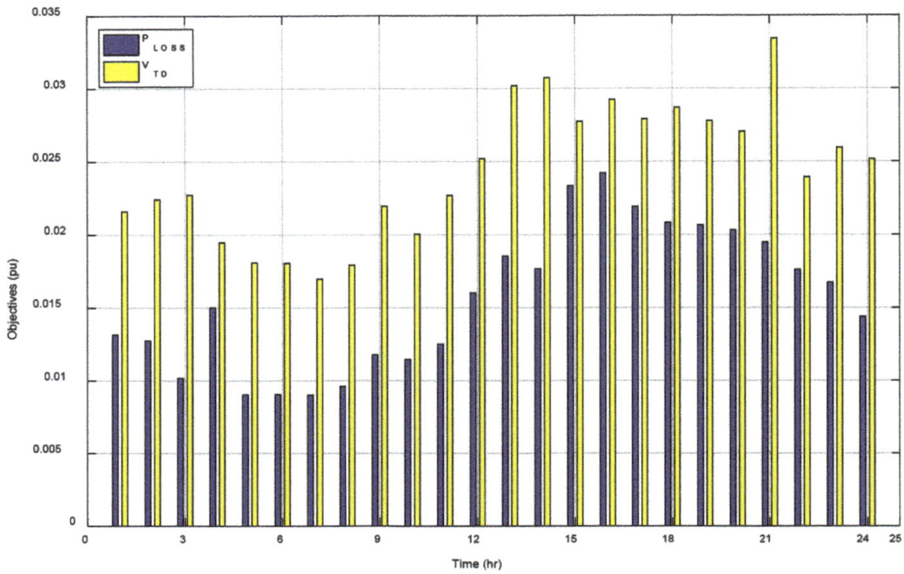

Figure 6.19 Best compromise solutions during the 24-hour operation using FDM for three OPF-DGs.

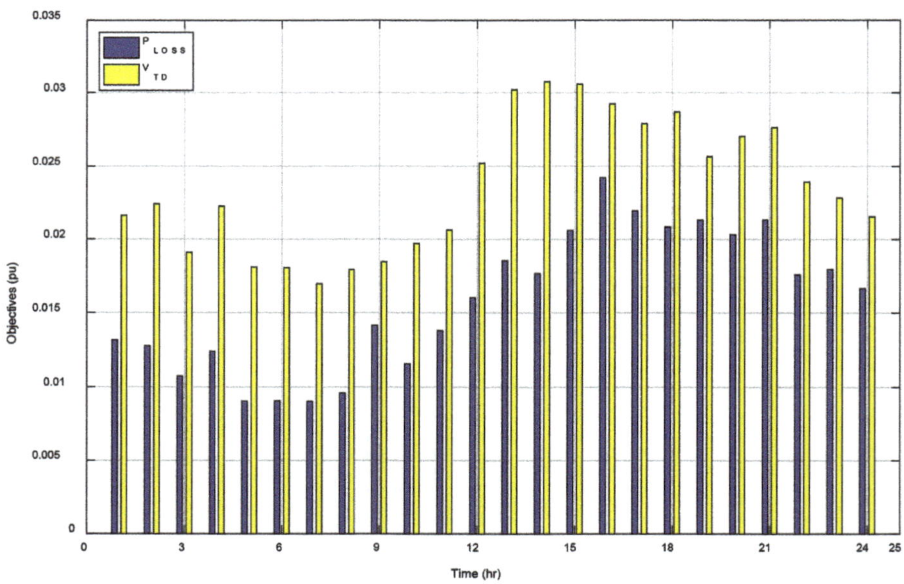

Figure 6.20 Best compromise solutions during the 24-holur operation using TOPSIS for three OPF-DGs.

Table 6.3 Average and standard deviation parameters for FDM and TOPSIS methods with three DGs.

DG PF	Parameter	FDM		TOPSIS	
		P_{LOSS}	V_{TD}	P_{LOSS}	V_{TD}
UPF	Ave	0.22515	0.09962	0.23713	0.05687
	Std	0.06363	0.01595	0.06657	0.00925
OPF	Ave	0.01564	0.02438	0.01584	0.02361
	Std	0.00482	0.00459	0.00471	0.00454

If we examine both graphs closely, we will notice that the power losses and voltage variations are less with OPF-DG than they were with UPF-DG.

The average and standard deviations of both objectives for UPF- and OPF-DG are listed in Table 6.3 for the case study of installing three DG units. The average and standard deviation for the OPF-DG are lower than the UPF-DG case. Moreover, more reductions are achieved compared to the corresponding single DG allocation.

6.6 Conclusion

Due to its adaptable, flexible, and decentralized paradigm, the 4th IR is more efficient and effective in the current market for renewable energy and power generation. Presently, there is a tremendous desire for energy utilities to boost their electricity generation from renewable energy sources to accomplish smart grid operation. Consequently, the advent of the 4th IR in power distribution systems has altered the energy market trend and the emergence of new technologies employing AI, IoT, and BD technologies for effective grid management. In the context of the 4th IR, the deployment of modern ICT communications offers various customer-focused and economically beneficial services. MOBO is utilized to optimize single and multiple DG units into distribution networks for both single and multiple objective optimization issues. Fuzzy-based decision-making (FDM) and TOPSIS approaches are utilized and contrasted for various MOO case studies. In most case studies, the results show that the TOPSIS technique outperforms the FDM method. Moreover, the voltage profile enhances with DG integration, and the system losses are decreased significantly with optimal PF DG units.

References

[1] L. Dogaru, "The main goals of the fourth industrial revolution. Renewable energy perspectives," *Procedia Manuf.*, vol. 46, pp. 397–401, 2020.

[2] R. Y. Zhong, X. Xu, E. Klotz, and S. T. Newman, "Intelligent Manufacturing in the Context of Industry 4.0: A Review," *Engineering*, vol. 3, no. 5, pp. 616–630, 2017.

[3] S. Wang, J. Wan, D. Zhang, D. Li, and C. Zhang, "Towards smart factory for industry 4.0: A self-organized multi-agent system with big data based feedback and coordination," *Comput. Networks*, vol. 101, pp. 158–168, 2016.

[4] M. Faheem *et al.*, "Smart grid communication and information technologies in the perspective of Industry 4.0: Opportunities and challenges," *Comput. Sci. Rev.*, vol. 30, pp. 1–30, 2018.

[5] X. Xu, Y. Lu, B. Vogel-Heuser, and L. Wang, "Industry 4.0 and Industry 5.0—Inception, conception and perception," *J. Manuf. Syst.*, vol. 61, no. September, pp. 530–535, 2021.

[6] S. Ing Tay *et al.*, "An Overview of Industry 4.0: Definition, Components, and Government Initiatives," *J. Adv. Res. Dyn. Control Syst.*, vol. 10, no. 14, pp. 1379–1387, 2018.

[7] S. Jayashree, M. N. H. Reza, C. A. N. Malarvizhi, and M. Mohiuddin, "Industry 4.0 implementation and Triple Bottom Line sustainability: An empirical study on small and medium manufacturing firms," *Heliyon*, vol. 7, no. 8, 2021.

[8] J. C. Kabugo, S. L. Jämsä-Jounela, R. Schiemann, and C. Binder, "Industry 4.0 based process data analytics platform: A waste-to-energy plant case study," *Int. J. Electr. Power Energy Syst.*, vol. 115, no. August 2019, 2020.

[9] M. Javaid, A. Haleem, R. P. Singh, R. Suman, and E. S. Gonzalez, "Understanding the adoption of Industry 4.0 technologies in improving environmental sustainability," *Sustain. Oper. Comput.*, vol. 3, no. January, pp. 203–217, 2022.

[10] S. Jayashree, M. N. H. Reza, C. A. N. Malarvizhi, A. Gunasekaran, and M. A. Rauf, "Testing an adoption model for Industry 4.0 and sustainability: A Malaysian scenario," *Sustain. Prod. Consum.*, vol. 31, pp. 313–330, 2022.

[11] A. Roozbeh Nia, A. Awasthi, and N. Bhuiyan, "Industry 4.0 and demand forecasting of the energy supply chain: A literature review," *Comput. Ind. Eng.*, vol. 154, no. October 2020, p. 107128, 2021.

[12] G. Beier *et al.*, "Impact of Industry 4.0 on corporate environmental sustainability: Comparing practitioners' perceptions from China, Brazil and Germany," *Sustain. Prod. Consum.*, vol. 31, pp. 287–300, 2022.

[13] S. Khan, S. Tomar, M. Fatima, and M. Z. Khan, "Impact of artificial intelligent and industry 4.0 based products on consumer behaviour

characteristics: A meta-analysis-based review," *Sustain. Oper. Comput.*, vol. 3, no. January, pp. 218–225, 2022.

[14] M. Chen, A. Sinha, K. Hu, and M. I. Shah, "Impact of technological innovation on energy efficiency in industry 4.0 era: Moderation of shadow economy in sustainable development," *Technol. Forecast. Soc. Change*, vol. 164, no. December 2020, p. 120521, 2021.

[15] Z. Shang, Z. He, Y. Chen, Y. Chen, and M. L. Xu, "Short-term wind speed forecasting system based on multivariate time series and multi-objective optimization," *Energy*, vol. 238, 2022.

[16] F. S. Tesch da Silva, C. A. da Costa, C. D. Paredes Crovato, and R. da Rosa Righi, "Looking at energy through the lens of Industry 4.0: A systematic literature review of concerns and challenges," *Comput. Ind. Eng.*, vol. 143, no. March, p. 106426, 2020.

[17] F. E. Boran, "A new approach for evaluation of renewable energy resources: A case of Turkey," *Energy Sources, Part B Econ. Plan. Policy*, vol. 13, no. 3, pp. 196–204, 2018.

[18] C. C. Sun, A. Hahn, and C. C. Liu, "Cyber security of a power grid: State-of-the-art," *Int. J. Electr. Power Energy Syst.*, vol. 99, no. November 2017, pp. 45–56, 2018.

[19] M. Kharrich *et al.*, "Optimal design of an isolated hybrid microgrid for enhanced deployment of renewable energy sources in Saudi Arabia," *Sustain.*, vol. 13, no. 9, pp. 1–26, 2021.

[20] A. Eid, S. Kamel, A. Korashy, and T. Khurshaid, "An Enhanced Artificial Ecosystem-Based Optimization for Optimal Allocation of Multiple Distributed Generations," *IEEE Access*, vol. 8, pp. 178493–178513, 2020.

[21] A. Eid, "Cost-based analysis and optimization of distributed generations and shunt capacitors incorporated into distribution systems with nonlinear demand modeling," *Expert Syst. Appl.*, vol. 198, p. 116844, Jul. 2022.

[22] M. Beltrami, G. Orzes, J. Sarkis, and M. Sartor, "Industry 4.0 and sustainability: Towards conceptualization and theory," *J. Clean. Prod.*, vol. 312, 2021.

[23] M. Ghobakhloo and M. Fathi, "Industry 4.0 and opportunities for energy sustainability," *J. Clean. Prod.*, vol. 295, p. 126427, 2021.

[24] M. Ghobakhloo, "Industry 4.0, digitization, and opportunities for sustainability," *J. Clean. Prod.*, vol. 252, p. 119869, 2020.

[25] T. Ahmad *et al.*, "Artificial intelligence in sustainable energy industry: Status Quo, challenges and opportunities," *J. Clean. Prod.*, vol. 289, p. 125834, 2021.

[26] S. K. Jha, J. Bilalovic, A. Jha, N. Patel, and H. Zhang, "Renewable energy: Present research and future scope of Artificial Intelligence," *Renew. Sustain. Energy Rev.*, vol. 77, no. May 2016, pp. 297–317, 2017.

[27] A. Maleki and A. Askarzadeh, "Comparative study of artificial intelligence techniques for sizing of a hydrogen-based stand-alone photovoltaic/wind hybrid system," *Int. J. Hydrogen Energy*, vol. 39, no. 19, pp. 9973–9984, 2014.

[28] W. Zhang, A. Maleki, F. Pourfayaz, and M. S. Shadloo, "An artificial intelligence approach to optimization of an off-grid hybrid wind/hydrogen system," *Int. J. Hydrogen Energy*, vol. 46, no. 24, pp. 12725–12738, 2021.

[29] P. Boza and T. Evgeniou, "Artificial intelligence to support the integration of variable renewable energy sources to the power system," *Appl. Energy*, vol. 290, no. February, p. 116754, 2021.

[30] S. M. Zahraee, M. Khalaji Assadi, and R. Saidur, "Application of Artificial Intelligence Methods for Hybrid Energy System Optimization," *Renew. Sustain. Energy Rev.*, vol. 66, pp. 617–630, 2016.

[31] A. S. Ali and B. J. Choi, "State-of-the-Art Artificial Intelligence Techniques for Distributed Smart Grids_ A Review.pdf," *Electron.*, vol. 9, no. 1030, pp. 1–25, 2020.

[32] M. A. M. Sadeeq and S. Zeebaree, "Energy Management for Internet of Things via Distributed Systems," *J. Appl. Sci. Technol. Trends*, vol. 2, no. 02, pp. 59–71, 2021.

[33] N. H. Motlagh, M. Mohammadrezaei, J. Hunt, and B. Zakeri, "Internet of things (IoT) and the energy sector," *Energies*, vol. 13, no. 2, pp. 1–27, 2020.

[34] T. Tan *et al.*, "Renewable energy harvesting and absorbing via multiscale metamaterial systems for Internet of things," *Appl. Energy*, vol. 254, no. March, p. 113717, 2019.

[35] C. Tipantuna, X. Hesselbach, and W. Unger, "Heuristic Strategies for NFV-Enabled Renewable and Non-Renewable Energy Management in the Future IoT World," *IEEE Access*, vol. 9, pp. 125000–125031, 2021.

[36] P. H. F. Sousa, N. M. M. Nascimento, J. S. Almeida, P. P. Rebouças Filho, and V. H. C. Albuquerque, "Intelligent Incipient Fault Detection in Wind Turbines based on Industrial IoT Environment," *J. Artif. Intell. Syst.*, vol. 1, no. 1, pp. 1–19, 2019.

[37] A. M. Eltamaly, M. A. Alotaibi, A. I. Alolah, and M. A. Ahmed, "Iot-based hybrid renewable energy system for smart campus," *Sustain.*, vol. 13, no. 15, pp. 1–18, 2021.

[38] P. D. Diamantoulakis, V. M. Kapinas, and G. K. Karagiannidis, "Big Data Analytics for Dynamic Energy Management in Smart Grids," *Big Data Res.*, vol. 2, no. 3, pp. 94–101, 2015.

[39] S. Mirjalili and J. S. Dong, *Multi-Objective Optimization using Artificial Intelligence Techniques*. Springer Nature Switzerland AG, 2020.

[40] S. Brisset and F. Gillon, *Approaches for multi-objective optimization in the ecodesign of electric systems*. Elsevier Ltd, 2015.

[41] N. Gunantara, "A review of multi-objective optimization: Methods and its applications," *Cogent Eng.*, vol. 5, no. 1, pp. 1–16, 2018.

[42] Y. Zhou-Kangas and K. Miettinen, "Decision making in multiobjective optimization problems under uncertainty: balancing between robustness and quality," *OR Spectr.*, vol. 41, no. 2, pp. 391–413, 2019.

[43] M. Soltanifar, "An investigation of the most common multi-objective optimization methods with propositions for improvement," *Decis. Anal. J.*, vol. 1, no. September, p. 100005, 2021.

[44] O. B. Augusto, F. Bennis, and S. Caro, "A new method for decision making in multi-objective optimization problems," *Pesqui. Operacional*, vol. 32, no. 2, pp. 331–369, 2012.

[45] A. Eid, S. Kamel, M. H. Hassan, and B. Khan, "A Comparison Study of Multi-Objective Bonobo Optimizers for Optimal Integration of Distributed Generation in Distribution Systems," *Front. Energy Res.*, vol. 10, no. 847495, pp. 1–14, 2022.

[46] A. A. A. Mohamed, S. Ali, S. Alkhalaf, T. Senjyu, and A. M. Hemeida, "Optimal allocation of hybrid renewable energy system by multi-objectivewater cycle algorithm," *Sustain.*, vol. 11, no. 23, pp. 1–20, 2019.

[47] M. H. Hassan, S. Kamel, L. Abualigah, and A. Eid, "Development and application of slime mould algorithm for optimal economic emission dispatch," *Expert Syst. Appl.*, vol. 182, no. April, p. 115205, 2021.

[48] A. Eid, "Performance improvement of active distribution systems using adaptive and exponential PSO algorithms," *Int. Rev. Electr. Eng.*, vol. 16, no. 2, pp. 147–157, 2021.

[49] H. Lotfi, R. Ghazi, and M. bagher Naghibi-Sistani, *Multi-objective dynamic distribution feeder reconfiguration along with capacitor allocation using a new hybrid evolutionary algorithm*, vol. 11, no. 3. Springer Berlin Heidelberg, 2020.

[50] H. Mohammadi, G. Khademi, M. Dehghani, and D. Simon, "Voltage stability assessment using multi-objective biogeography-based subset selection," *Int. J. Electr. Power Energy Syst.*, vol. 103, no. April, pp. 525–536, 2018.

[51] A. Zeinalzadeh, Y. Mohammadi, and M. H. Moradi, "Optimal multi objective placement and sizing of multiple DGs and shunt capacitor banks simultaneously considering load uncertainty via MOPSO approach," *Int. J. Electr. Power Energy Syst.*, vol. 67, pp. 336–349, 2015.

[52] A. Eid, "Allocation of distributed generations in radial distribution systems using adaptive PSO and modified GSA multi-objective optimizations," *Alexandria Eng. J.*, vol. 59, no. 6, pp. 4771–4786, 2020.

[53] S. Rezaeian Marjani, V. Talavat, and S. Galvani, "Optimal allocation of D-STATCOM and reconfiguration in radial distribution network using MOPSO algorithm in TOPSIS framework," *Int. Trans. Electr. Energy Syst.*, vol. 29, no. 2, pp. 1–25, 2019.

[54] B. Sun, "A multi-objective optimization model for fast electric vehicle charging stations with wind, PV power and energy storage," *J. Clean. Prod.*, vol. 288, p. 125564, 2021.

[55] S. Abdel Aleem, A. Y. Almoataz, A. F. Zobaa, and R. C. Bansal, *Decision Making Applications in Modern Power Systems*, 1st ed. London: Academic Press, 2020.

[56] H. Falsafi, A. Zakariazadeh, and S. Jadid, "The role of demand response in single and multi-objective wind-thermal generation scheduling: A stochastic programming," *Energy*, vol. 64, pp. 853–867, 2014.

[57] S. Chakraborty, "TOPSIS and Modified TOPSIS: A comparative analysis," *Decis. Anal. J.*, vol. 2, no. December 2021, p. 100021, 2022.

[58] A. K. Das, A. K. Nikum, S. V. Krishnan, and D. K. Pratihar, "Multi-objective Bonobo Optimizer (MOBO): an intelligent heuristic for multi-criteria optimization," *Knowl. Inf. Syst.*, vol. 62, no. 11, pp. 4407–4444, Nov. 2020.

[59] K. Deb, A. Pratap, S. Agarwal, and T. Meyarivan, "A Fast and Elitist Multiobjective Genetic Algorithm: NSGA-II," 2002.

[60] R. Ranjan, B. Venkatesh, and D. Das, "Voltage stability analysis of radial distribution networks," *Electr. Power Components Syst.*, vol. 31, no. 5, pp. 501–511, 2003.

[61] S. F. Zandrazavi, C. P. Guzman, A. T. Pozos, J. Quiros-Tortos, and J. F. Franco, "Stochastic multi-objective optimal energy management of grid-connected unbalanced microgrids with renewable energy generation and plug-in electric vehicles," *Energy*, vol. 241, no. 122884, pp. 1–17, 2022.

Appendix

The complete data of the 69-bus distribution system are listed in Table A1

Table A1 69-bus system data and demand.

	Impedance		Demand			Impedance		Demand	
Branch	r (Ω)	x (Ω)	P (kW)	Q (kVar)	Branch	r (Ω)	x (Ω)	P (kW)	Q (kVar)
1	0.0005	0.0012	0	0	35	0.0044	0.0108	26	18.55
2	0.0005	0.0012	0	0	36	0.0640	0.1565	26	18.55
3	0.0015	0.0036	0	0	37	0.1053	0.1230	0	0
4	0.0251	0.0294	0	0	38	0.0304	0.0355	24	17
5	0.3660	0.1864	2.6	2.2	39	0.0018	0.0021	24	17
6	0.3810	0.1941	40.4	30	40	0.7283	0.8509	1.2	1
7	0.0922	0.0470	75	54	41	0.3100	0.3623	0	0
8	0.0493	0.0251	30	22	42	0.0410	0.0478	6	4.3
9	0.8190	0.2707	28	19	43	0.0092	0.0116	0	0
10	0.1872	0.0619	145	104	44	0.1089	0.1373	39.22	26.3
11	0.7114	0.2351	145	104	45	0.0009	0.0012	39.22	26.3
12	1.0300	0.3400	8	5	46	0.0034	0.0084	0	0
13	1.0440	0.3450	8	5.5	47	0.0851	0.2083	79	56.4
14	1.0580	0.3496	0	0	48	0.2898	0.7091	384.7	274.5

(Continued)

Table A1 *Continued.*

Branch	Impedance		Demand		Branch	Impedance		Demand	
	r (Ω)	x (Ω)	P (kW)	Q (kVar)		r (Ω)	x (Ω)	P (kW)	Q (kVar)
15	0.1966	0.0650	45.5	30	49	0.0822	0.2011	384.7	274.5
16	0.3744	0.1238	60	35	50	0.0928	0.0473	40.5	28.3
17	0.0047	0.0016	60	35	51	0.3319	0.1114	3.6	2.7
18	0.3276	0.1083	0	0	52	0.1740	0.0886	4.35	3.5
19	0.2106	0.0690	1	0.6	53	0.2030	0.1034	26.4	19
20	0.3416	0.1129	114	81	54	0.2842	0.1447	24	17.2
21	0.0140	0.0046	5	3.5	55	0.2813	0.1433	0	0
22	0.1591	0.0526	0	0	56	1.5900	0.5337	0	0
23	0.3463	0.1145	28	20	57	0.7837	0.2630	0	0
24	0.7488	0.2475	0	0	58	0.3042	0.1006	100	72
25	0.3089	0.1021	14	10	59	0.3861	0.1172	0	0
26	0.1732	0.0572	14	10	60	0.5075	0.2585	1244	888
27	0.0044	0.0108	26	18.6	61	0.0974	0.0496	32	23
28	0.0640	0.1565	26	18.6	62	0.1450	0.0738	0	0
29	0.3978	0.1315	0	0	63	0.7105	0.3619	227	162
30	0.0702	0.0232	0	0	64	1.0410	0.5302	59	42
31	0.3510	0.1160	0	0	65	0.2012	0.0611	18	13
32	0.8390	0.2816	14	10	66	0.0047	0.0014	18	13
33	1.7080	0.5646	9.5	14	67	0.7394	0.2444	28	20
34	1.4740	0.4873	6	4	68	0.0047	0.0016	28	20

Index

About the Editors

Prof. Diaa-Eldin A. Mansour was born in Tanta, Egypt in 1978. He received his B.Sc. and M.Sc. degrees in Electrical Engineering from Tanta University, Tanta, Egypt, in 2000 and 2004, respectively, and his Ph.D. degree in Electrical Engineering from Nagoya University, Nagoya, Japan, in 2010. Since 2000, he has been with the Department of Electrical Power and Machines Engineering, Faculty of Engineering, Tanta University, Egypt, where he has been a professor since 2020. He is currently working as a professor in Electrical Power Engineering Department, Egypt-Japan University of Science and Technology (E-JUST), Alexandria, Egypt, on leave from Tanta University. His research interests include high voltage engineering, condition monitoring of electrical power equipment, IoT applications in electrical power systems, and applied superconductivity. He twice received the Best Presentation Award from IEE of Japan in 2008 and 2009, the Prof. Khalifa Prize from the Egyptian Academy of Scientific Research and Technology in 2013, the Tanta University Encouragement Award in 2016, the Egypt-State Encouragement Award in the field of engineering sciences in 2018, and the Tanta University Citations Award in 2021. He was listed among the world's top 2% of scientists by Stanford University, USA, from 2020 to 2023.

Dr. Mohamed Gamal Hussien received his B.Sc. and M.Sc. degrees from the Department of Electrical Power and Machines Engineering, Faculty of Engineering, Tanta University, Egypt in 2011 and 2016, respectively, in Electrical Engineering. In 2020, he received his Ph.D. degree from Huazhong University of Science and Technology, China. Dr. Hussien is currently an Assistant Professor at the Department of Electrical Power and Machines Engineering, Faculty of Engineering, Tanta University, Tanta, Egypt. He has published scientific papers in both international conferences and high-quality SCI journals. He is currently an Associate Editor for the journal IET Renewable Power Generation. He is also currently an Academic Editor for the International Transactions on Electrical Energy Systems journal. He serves as a reviewer for IEEE conferences and many IEEE transactions journals.

He is a member of the IEEE-IES Electric Machines Technical Committee and a member of the IEEE-IES Technical Committee on Renewable Energy Systems. His paper has been selected as one of the best papers published in the period 2019–2020 in IEEE Transactions on Energy Conversion in the "power generation systems and grid interfaces" area. He has also received the award of Best Paper from the IEEE IAS GUCON 2021 conference.

Dr. Partha Sarathi Subudhi completed his B.Tech. in Electrical Engineering from Biju Patnaik University of Technology, Odisha, India in 2012 and M.Tech. in Power Electronics and Drives from Vellore Institute of Technology, Chennai, India in 2015. He gained his Ph.D. with the School of Electrical Engineering (SELECT), Vellore Institute of Technology (VIT), 2015–2020. He also served as a Teaching cum Research Assistant at the same time. Currently, he is an Assistant Professor with the Department of Electrical Engineering, Bajaj Institute of Technology, Wardha, Maharashtra, India. He is an Editor of International Journal of Power and Energy Systems, and Editor of DGAEJ, River Publishers. Dr. Subudhi is a senior member of IEEE, a member of the IEEE Technical Committee (TC 9) on Wireless Power Transfer Systems, a member of IEEE Technical Committee (TC 12) on Energy Access and Off-Grid System, a member of the IEEE PELS Educational Videos Committee, and a life member of ISTE and IEANG. His fields of interest include wireless power transfer, electric vehicle charger, residential nano grid, solar power generation system, hybrid converters, and their applications.

Dr. A. Rakesh Kumar completed his B.Eng. with honors in Electrical and Electronics Engineering from Anna University, Chennai, India in 2011 and M.Eng. in Power Electronics and Drives from Anna University, Chennai, India in 2013. He worked as Assistant Professor with the Department of EEE, Rajalakshmi Engineering College, Chennai, India from 2013 to 2015. He then went on to a full-time PhD with the School of Electrical Engineering (SELECT), Vellore Institute of Technology (VIT) from 2015 to 2019. He also served as Teaching cum Research Assistant at the same time. Subsequently, he completed his Post-Doctoral Fellowship from National Institute of Technology Tiruchirappalli, India. He is an Assistant Professor in the Department of EEE, KPR Institute of Engineering and Technology, India. He is an IEEE Senior member and he is actively engaged with the IEEE Power Electronics Society. He is also a member of the IEEE PELS Educational Videos Committee and the Managing Editor of IEEE Educational Videos on

Power Electronics. His fields of interest include multilevel inverters, inverter modulation techniques, and nanogrids and their applications.

Prof. Sanjeevikumar Padmanaban is a Full Professor in Electrical Power Engineering with the Department of Electrical Engineering, Information Technology, and Cybernetics, University of South-Eastern Norway, Norway. He has authored over 750+ scientific papers and received Fellow of the Institution of Engineers, India, the Institution of Electronics and Telecommunication Engineers, India, and the Institution of Engineering and Technology, UK. He received a lifetime achievement award from Marquis Who's Who, USA 2017 for contributing to power electronics and renewable energy research. He is listed among the world's top 2% of scientists (from 2019) by Stanford University USA.

For Product Safety Concerns and Information please contact our EU representative GPSR@taylorandfrancis.com
Taylor & Francis Verlag GmbH, Kaufingerstraße 24, 80331 München, Germany

www.ingramcontent.com/pod-product-compliance
Ingram Content Group UK Ltd.
Pitfield, Milton Keynes, MK11 3LW, UK
UKHW022211190926
475702UK00003B/304
9788770226769